Views of Science, Technology and Development

Views of Science, Technology and Development

Edited by

†EUGENE RABINOWITCH and VICTOR RABINOWITCH

PERGAMON PRESS

OXFORD · NEW YORK · TORONTO
SYDNEY · PARIS · BRAUNSCHWEIG

U.K.	Pergamon Press Ltd., Headington Hill Hall, Oxford OX3 0BW, England
U.S.A.	Pergamon Press Inc., Maxwell House, Fairview Park, Elmsford, New York 10523, U.S.A.
CANADA	Pergamon of Canada, Ltd., 207 Queen's Quay West, Toronto 1, Canada
AUSTRALIA	Pergamon Press (Aust.) Pty. Ltd., 19a Boundary Street, Rushcutters Bay, N.S.W. 2011, Australia
FRANCE	Pergamon Press SARL, 24 rue des Ecoles, 75240 Paris, Cedex 05, France
WEST GERMANY	Pergamon Press GmbH, 3300 Braunschweig, Postfach 2923, Burgplatz 1, West Germany

---

First edition 1975

Library of Congress Cataloging in Publication Data

Main entry under title:

Views of Science, technology, and development.

Consists of papers taken from recent Pugwash Conferences.

1. Science—International cooperation—Congresses. 2. Technology—International cooperation—Congresses. 3. Underdeveloped areas—Science—Congresses. 4. Underdeveloped areas—Research—Congresses. 5. Technical assistance—Congresses. I. Rabinowitch, Victor, ed. II. Rabinowitch, Eugene I., 1901–1973 ed. III. Pugwash Conference on Science and World Affairs. Q172.5.I5V5 1975 338.9′009172′4 74-32201

ISBN 0-08-018241-0

Printed in Great Britain by A. Wheaton & Co., Exeter

Contents

CHAPTER I

Introduction

Introduction

IN RECENT years there has been a growing appreciation of the importance of science and technology in influencing the conditions of human existence. This is no less true in the so-called developing or "third" world than in the affluent developed world. In the latter the main concern is that of *controlling* science and technology for the benefit of society while in the former the major concern is that of *harnessing* science and technology for the economic and social development of society. Responding to both of these concerns requires that traditional patterns of individual and group behavior developed through biological and social evolution be changed to meet these new conditions of human existence. What these changes must be and how they can be brought about without creating social chaos are among the major questions facing the world's peoples today.

Never before in man's history has the potential existed as it does now for the provision to all the world's people of the requirements of the "good life"—this largely due to the results of the "Scientific and Technological Revolution". Indeed economists tell us that at the present time, an "average human being" on earth has at his disposal the equivalent of a dozen able-bodied slaves to do his bidding. And yet it is being increasingly suggested by some thoughtful people that it is time to slow down or stop altogether ever-accelerating growth. Otherwise, we are warned, the combined effects of exhaustion of natural resources, pollution of our habitat on earth, and exploding population will combine to bring about an "overshoot" followed by catastrophic collapse.

Others answer that the three ills—exhaustion of resources, pollution of the environment and population explosion—are susceptible to "technological treatment", provided we have enough energy to invest in energy-intensive undertakings such as liberation of critical materials from low-grade sources (sea water, common rocks), recycling of used materials, abatement of pollution, and repair of damage inflicted on our habitat by the industrial revolution and population explosion.

Still others warn that apart from limitations imposed on energy production by exhaustion of fossil fuels—which might be overcome through development of fission breeders and fusion power—a rigid ceiling is imposed on increasing energy production by the danger of upsetting the "metastable meteorological equilibrium" prevailing on earth since the last ice age and provoking catastrophic meteorological effects, above all the melting of polar ice caps.

In sum, as a result of all of these considerations and others, we are urgently being advised to slow down world-wide growth, which implies renunciation of energy-consuming technologies in favor of the utilization of abundant natural materials, recycling, and abatement of pollution and other forms of environmental abuse.

Were the world confined to Europe and North America plus perhaps Japan, South Korea, Taiwan and Brazil—the so-called "developed" or "near developed" countries—the advice given us might be considered bitter but not tragic. For the rest of the world the so-called "Third World" acceptance of such advice flies in the face of aspirations for better health, freedom from hunger, and increased opportunities generally for a better life—all of which have been promised through the technological revolution in the more developed parts of the world and communicated through every mode of communication from word of mouth to satellite communications.

It is in part this dilemma which provides the great challenge to mankind in the next decade. Can one-third of civilization cope in a world of decreasing natural resources and increased population growth, while the other two-thirds of the world achieve its greatest aspirations for increased well-being and even affluence?

It is a concern for the role of science and technology in contributing significantly to answering this question to which this volume is devoted. It is our firm belief that only through the thoughtful and careful application of science and technology, with due consideration for their relationship to social and cultural aspirations, can effective development without catastrophic consequences take place.

The papers included in this volume have been largely, though not exclusively, taken from recent conferences on science and world affairs—the "Pugwash Conferences", in which scientists, both natural and social, have focused their attention on the two major problems facing humanity, world peace and development.

Though several of these conference papers were prepared some time ago, they have in almost all cases been brought up to date very recently. Others are included in order to provide a historical reference thought by the editors to be of particular significance.

We are grateful to all authors for their patience in awaiting final publication of their papers. The delay is the fault solely of the editors who as a result of changes in residence and position found it impossible to devote the necessary time to complete the task.

We would like to express our gratitude to Mr. Warren Hall, Miss Rose Ameser, and Ms. Donna Shipley for their assistance in various stages of the preparation of this book. Thanks are also due to the Pugwash Conferences on Science and World Affairs for making available to the editors the papers from their conferences and symposia.

Finally, I cannot complete this introduction without acknowledging the primary role of my father in conceiving the idea for this volume and in making the initial contacts with contributing authors. It was his deep commitment to science and technology and his faith that they could provide the basis for a better life for all humanity that made this collection one of his primary concerns up to his death in May 1973.

Washington, D.C.
February 2, 1974

VICTOR RABINOWITCH

CHAPTER II

Views from Developing Countries

*Brain-drain from Developing Countries**

ASHOK PARTHASARATHI†

THERE are several reasons for developing countries to continue to be concerned about their loss of scientific and technical manpower through migration. First, there is increasing evidence that some developing countries are losing so much of their manpower that their capacity to tackle their social, economic and cultural problems is being damaged irreparably. Second, much international cooperation is based on the idea that the high-income countries should be supporters rather than prime movers. But this can be a meaningful basis for aid only if developing countries possess their own primary scientific resources. The magnitude and quality of the talent that many developing countries will lose if present trends continue, however, raise serious doubts about the generation and growth of sufficient endogenous scientific and technological activity for any external action to be of much use. Third, the response of the high-income countries to this problem is an excellent indication of whether they are prepared to place commitment to world-wide development above short-term national self-interest.

My suggestions for policies and action which can and must be taken now on a unilateral, bilateral and multilateral basis, are based on at least two premises. First, statistics of the emigration of scientific and technical personnel from low- to high-income countries, and the composition and dynamics of such emigration during the past 8 years, indicate that some categories of manpower at least constitute a loss to the country concerned. Second, there is evidence that the total outflow of students, trainees and "fully trained manpower" from low- to high-income countries is increasing rapidly.

Responsibilities at Home

Certain initiatives should be taken by the developing country itself. The Government of India, for example, has progressively reduced the range of disciplines and areas of specialization for which an Indian can go abroad to study. In such cases, before permission is given, students must show that they have reached the highest level of training available in India. It should, however, be a general policy that no overseas study is allowed until the student or trainee has been educated to the highest

* This paper first appeared in *Nature*, **230**, 87–90 (1971).

† Prime Minister's Secretariat, New Delhi, India.

possible level in his own country. After this every student or trainee should be employed at home for at least 2 years, before being permitted to study abroad. Such a requirement, which of course needs legal sanction, must also be backed by several governmental and non-governmental measures to make use of the students.

Available information about apprentice-training programmes in India suggests that private industry tries to avoid taking its share of the social cost of such training. Governments would therefore have to take such measures as instituting industrial fellowship schemes for supporting engineering manpower in industry for the 2 years. Similar support would have to be provided in medical and other government services, and also in major research and development establishments. Such a procedure might, in an appropriate political and economic environment, familiarize the young scientist, engineer or doctor with real developmental tasks. This experience could have a powerful effect on the discipline or area that he chooses for further study abroad (at the end of the 2 years) or may even lead to a re-evaluation of the costs and benefits of overseas study.

It is appropriate and necessary to ask whether developing countries, such as India and the United Arab Republic (Egypt), can continue to run a "mixed economy" in their manpower development programme. Is it meaningful to bring domestic higher education within the purview of planning, while leaving overseas training to the "international manpower market"? It should be remembered that the forces determining the dynamics of that market are beyond the control of the developing countries concerned. It seems, therefore, that it will be impossible to avoid the setting of quotas for overseas training in terms of countries, type of degree/training programme, sub-disciplines and the intellectual attainment of candidates, if meaningful manpower planning is to be undertaken. There have been precedents for this approach, from Japan in the 1880s to the technical-aid programmes of the communist countries today, in which training is imparted according to disciplines, sub-disciplines and even specific skills, which are needed in relation to committed programmes.

Maintaining Communication

Systematic opinion surveys as well as observation and experience suggest that when specialists from developing countries are abroad they lose contact with domestic, professional, social, economic and political developments to a high degree. For example, a pilot survey carried out in 1967 in the Greater Boston area in the United States revealed that 90% of Indian students of natural science and technology never saw the bulletins of leading research organizations such as the Council of Scientific and Industrial Research, the Department of Atomic Energy or the University Grants Commission. Foreign student advisers in American universities have indicated that a special government publication, which details opportunities in research and education in India and which is sent to all major American and European universities, is seldom seen by Indian students on their campuses. Even weekly editions of Indian newspapers are seen infrequently.

And so it is perhaps not surprising that India's overseas professionals are unaware of major initiatives being taken by the Government of India to create more opportunities for skilled professionals in, for example, the electronics and aeronautics industries, small-scale manufacturing and agricultural research and development. They are continuously exposed to their social, economic and professional developments in their country of study and simultaneously to partial and often slanted reports on India by the mass media of their host country. This converts the information gap into one of cynicism and despondency about the possibility of India's progress in general and, in particular, their own role in it.

To combat this situation, it is imperative that, first, the overseas circulation of the bulletins of research and educational institutions of the developing countries is expanded, with a view to reaching individual students. Second, educational and cultural attachés in the embassies of developing countries in the United States, United Kingdom and Canada should periodically visit universities with large concentrations of their students, not merely to provide information about jobs at home, but to help to solve problems. In this way trust and confidence will be built up. Third, leading scientists and administrators from developing countries should visit major university campuses when they are in North America and Europe. On such occasions the focus should be on scientific and technological programmes at home, trends in scientific, technological and educational policy and so on.

Before they leave their home countries all students and self-sponsored trainees should be required to undertake a "familiarization programme", for which the following model might be considered.

Students who have secured admission to a foreign university could be brought together for discussions and seminars on current and projected scientific and technological problems and programmes, led by outstanding practitioners from various disciplines. Such meetings would also put the role of the professional in the context of the overall developmental effort of the home country. After a plenary or general session of this kind, during which the seminar leaders would live with the students to ensure informal contact, the students could be broken up into small groups on a disciplinary basis and "teach-ins" could be conducted by specialists. The potential foreign trainees would then be taken around some of the country's major industrial, agricultural, research and educational centres to see the day to day problems of development. The whole programme might then culminate in a meeting with the prime minister or president of the country, at which stage some exhortation may be introduced. Such a scheme might seem far too ambitious and expensive. But India, for example, has been running such a programme, for civil service trainees, for more than 10 years. The nature and aims of the effort will be different, but it ought to be found to be at least as effective.

Recruiting Foreign Professionals

So far the Government of India, for example, has relied almost entirely on the Scientists Pool Programme to enable foreign professionals to be recruited by Indian organizations. There have been indications for some time, however, that this will not suffice. Direct overseas recruitment by employers is necessary. The details of any scheme to undertake such recruitment will vary from country to country, but any such effort should involve at least the preparation of a list indicating all budgeted, but unfilled, vacancies for professionals in laboratories, government departments and universities; there should also be detailed job descriptions in each case, and the maximum emoluments the employer would be prepared to pay to a professional meeting his requirements. A selection panel of professionals, from industry, universities and research institutions, who are not only "senior" enough but whose disciplinary/

RATIO OF THE ANNUAL NUMBER OF EMIGRANTS TO THE ANNUAL NUMBER OF GRADUATES FROM THE RELATED FIELD OF STUDY BY COUNTRY OF ORIGIN OR LAST PERMANENT RESIDENCE OVER THE PERIOD 1962–6

	Engineers		Natural scientists		Physicians		Nurses		Social scientists
	France and United States	France and United States and Canada	France and United States	France and United States and Canada	France and United States	France and United States and Canada	United States	United States and Canada	United States
Europe									
Greece	23.9	26.6	10.3	12.2	5.9	7.7	5.1	8.3	0.2
Turkey	6.5	6.9	12.6	15.9	9.8	15.2	1.9	2.8	0.1
Asia									
Burma	N.a.	N.a.	N.a.	N.a.	1.3	2.9	0.7	1.0	N.a.
Hong Kong	64.2	514.2	31.6	59.3	22.2	65.1	N.a.	N.a.	12.3
India	4.4	6.1	N.a.	N.a.	0.6	1.6	0.1	1.1	0.1
Indonesia	N.a.	N.a.	N.a.	N.a.	—	0.1	N.a.	N.a.	N.a.
Iran	47.5	51.9	12.3	14.1	7.7	10.0	3.3	3.9	∞
Iraq	7.5	9.2	N.a.	N.a.	4.3	4.9	N.a.	N.a.	0.1
Israel	12.7	15.7	15.8	18.6	43.8	54.3	9.4	11.4	N.a.
Lebanon	33.1	35.5	9.4	10.5	15.2	24.9	N.a.		0.5
Malaysia	N.a.	N.a.	N.a.	N.a.	N.a.	N.a.	N.a.		N.a.
Pakistan	3.2	9.2	0.2	0.5	0.5	1.4	N.a.		0.01
Philippines	0.4	2.3	3.6	18.5	12.0	19.3	3.6	35.3	0.03
Republic of Korea	4.0		2.4		3.1		1.1		1.5
Republic of Viet-Nam	76.3		17.6		1.0		N.a.		N.a.
Syria	54.2	56.5	10.9	11.7	6.7	9.3	N.a.		0.8
Thailand	1.0		0.1		2.9		2.6		0.04

This table is taken from the Official Records of the UN General Assembly twenty-third session 1968. Annexes to Agenda item 47.

technological specializations broadly match that of the people to be recruited, should then be sent abroad on a recruiting mission. Apart from its portfolio of "professionals needed and funded", it is imperative that the panel has the authority to offer appointments and emoluments to suitable interviewees immediately. If a developing country does not have the necessary mature professionals to staff such panels, United Nations agencies may be requested to provide them as consultants.

A major stumbling block in the return of fully trained professionals to many developing countries has been the fact that they have little choice but to join an organization in their own country. This means that they have to conform to the behavioural and intellectual environments of such organizations, the poverty and rigidity of many of which are deep. It therefore seems necessary to design institutional structures in which overseas professionals would have greater control over their own activities. A design and development estate and a co-operative hospital are two examples that might be considered.

The design and development estate would have many similarities with the industrial estates that already exist in several developing countries—an industrial development corporation would provide land and certain facilities and hire out laboratory and engineering space to individual engineers or partnerships. Engineers returning home would also need low-interest loans and foreign-exchange allocations. They could then use these resources and their skills to take on contract work for government and other agencies.

The co-operative hospital would have some similarities with the development estate, but would clearly require more planning and organization and also financial aid from the government. A group of medical doctors and hospital administrators could be asked to examine how such a project might be tackled operationally.

Responsibilities Abroad

It is widely accepted in scientific and academic circles in the United States and the United Kingdom—the principal destinations of professional manpower from the developing nations—that the present immigration policies of these countries are discriminatory, being heavily biased in favour of high-level manpower. This situation needs urgent redress if there is to be any credibility behind the protestations that these countries are willing to implement all non-punitive measures to reduce and reverse the loss of talent from developing countries. Operational measures are not difficult to devise. For example, in the United States, three initiatives are immediately possible. First, if the United States wishes to limit the total flow of immigrants, the entry of that number of immigrants should be entirely on a first-come first-served basis. Second, after completing their training or work all holders of "exchange visitor" visas should be required to return to their home country for a certain period before applying to enter the United States as immigrants. Third, the specified period of stay at home should be increased from 2 to 4 years. The U.S.

Government should then indicate its willingness to apply these changed regulations to those students and other professionals whose national governments make a specific request to this effect. The regulations need not be made to apply to all categories of professionals, but only to those such as medical doctors, who are specifically indicated by the developing country concerned.

In case of some high-income countries, an inadequate domestic supply of certain types of manpower is directly responsible for the loss of brains from developing countries. For the past 7 or 8 years, 25 to 30 per cent of all residents and interns in U.S. hospitals have been from the developing countries. Yet, under pressure from such powerful lobbies as the American Medical Association, the intake of local students into U.S. medical colleges and universities continues to be carefully limited. Consequently, a large unfulfilled demand for medical services has built up and U.S. hospitals are meeting it by consciously depending on a steady inflow of freshly trained doctors from the developing countries.

It has often been suggested that countries like the United States and the United Kingdom should have, at least in a few universities, courses in natural science and engineering oriented towards the problems of developing countries. But the fact that the level of specialization is often as important as the content and orientation of courses seems to have received less attention. This aspect is important, however, because there is evidence that the probability of a developing country "losing" one of its students abroad is related to whether his study programme involves a first, a second or a third degree. My investigations have revealed that for 300 Indian alumni of the "top ten" American universities, during 1956–66, the probability of immediate return home, after taking a first degree was 50%, after taking a second degree it was 41% and after gaining a Ph.D. it was only 18%. The variance in each case was close to 5%. Similar results have also been obtained by other investigators (private communication from D. D. Henry and C. Susskind).

Although many factors must be taken into account in interpreting these figures, the degree of specialization and probability of return do seem to be related, which is hardly surprising in view of the following arguments. First, although the marginal economic advantage to a first-degree holder in acquiring a second degree is not very different in a highly industrialized country from that in a developing country, the differential advantage is significant if he acquires a third degree, because of the nature of the two economies. Thus, when a student from a developing country completes his Ph.D. and then assesses his "comparative market value" in his home country and in the country in which he has studied, the differential in favour of the latter is so great that his choice of whether or not to return home is virtually pre-empted. This, however, is not so in the case of a student who has just acquired a master's degree. Second, educationalists agree that, in most industrialized countries, the Ph.D. involves such intense specialization, at such an impressionable stage of the student's intellectual and cultural development, that his inclinations are heavily oriented, not only towards an academic or research career, but towards one in which he will be able to continue to work in the narrow sub-speciality which he chose as his thesis topic.[1] Such a person has alienated himself from his home country without giving it a chance to secure and hold his interest and commitment.

Education Overseas

It would thus seem necessary for more students to be urged and, if necessary, required by their own governments to terminate their overseas education after the second degree. The difficulty is that in many universities in the highly industrialized countries, master's degree programmes no longer exist. Thus the governments of developing countries can take the necessary administrative steps only if some of the leading universities in each high-income country, and particularly the United States, start 1- and 2-year master's degree courses. These, together with the resolve of developing countries to allow, as far as possible, only first-degree holders to study abroad, would also prevent too great an acculturation of the student into the ways of his host country, which seems considerably to influence his decision whether or not to return home. This, however, raises the question of whether preventing such acculturation conflicts with the foreign policy dividends which most industrialized countries expect from their educational and cultural exchange programmes.

It is the responsibility of the high-income countries to assist the developing countries to recruit the latter's professionals who are currently abroad. The kind of assistance needed will depend on the countries involved, but a general problem is to locate engineers in industry and, to some extent, medical doctors whether in service or research. The departments of education of the high-income countries should be asked to help the embassies of the developing countries in such location. The governments of the high-income countries should also allow their aid funds to be used to support official efforts by a developing country to recruit professionals so located. The annual visits of the selection panels I mentioned earlier might be a case in point.

Joint Initiatives

Although many different kinds of measures can obviously be devised within the framework of existing bilateral programmes between less and more developed countries, I shall take a specific Indian example to demonstrate the feasibility of such measures. In 1963 the University Grants Commission of the Government of India started a Summer Institutes Programme in collaboration with the United States National Science Foundation and the Agency for International Development of the United States. The purpose of the programme was to run 8-week courses for school and college teachers of the natural sciences, to familiarize them with new curricula, teaching methods, and experimental approaches. The institutes were organized on a disciplinary basis and manned by faculty members from American universities, whose numbers have grown from 12 to 179 during the past 5 years.[2] The programme offers an excellent opportunity for Indian professionals currently holding faculty positions in American universities—of which there were 925 when last counted in

1968[3]—to pay a short visit to their home country. Such visits would enable at least the potential "brain-drainee" to renew family ties, visit Indian research, educational and industrial institutions, interact with professional colleagues there, survey employment opportunities and generally become aware at first hand of the challenges (and hazards) of returning for good. Although this initiative is basically joint in nature, the first move must be taken by the United States Government, if for no other reason than that American aid funds are being used to cover a significant part of the cost of this programme.

Student counsellors are needed for postgraduate students from developing countries who are studying abroad. Something must be done to orientate the postgraduate student from a developing country, ideally throughout his programme but at least when he begins to think about the topic for his thesis. Any such scheme requires considerable cooperation on the part of the foreign university, and so a start should probably be made with one country, the United States. The mechanics of the scheme, again using India as an illustration, might be as follows. About 100 students would be selected from those admitted to postgraduate natural science and engineering programmes for the coming academic year, on the basis of their academic quality and, to some extent, the broad relation between the disciplines in which they intend to work and national needs. About twenty counsellors in the same disciplines would be appointed from university faculty, staff of research laboratories and professionals in industry. Each would be "assigned" five students for whom he would act as "home country interface" on both professional and social matters. Students would have several meetings with their counsellors even before leaving home. (This might be usefully combined with the familiarization programme discussed earlier.)

Soon after the student had gone to the United States, the counsellor would contact the chairman of the university department where his particular students had been registered to discuss the student's background and interests, courses, research programmes and so on. He would also keep in close touch with his students and do his best to influence their choice of course consistent with the department's programme. During the course the counsellor would provide professional advice and keep him informed of problems at home which illustrate the principles and techniques that the student is learning. As the student approached the time to begin his thesis the counsellor would discuss with the American university the range of research topics available, and their relation to the aptitude of his student and to Indian needs. Eventually several topics would be chosen by the Indian counsellor and the American supervisor for each student to select from. The student would, of course, be kept in the picture.

Such a scheme has two essential merits. First, it gives the student a personal and institutional link with his home country. Second, it provides continuing evidence of its stake in the direction his studies take. But it would not be easy to implement such a scheme.

International corporations have an important part to play in assisting developing country professionals to return home. Foreign governments and commercial companies have to recognize the need for the middle and top management of foreign firms operating in developing countries to be recruited from local nationals. It should not be difficult therefore to encourage foreign governments to urge that companies operating in developing countries recruit local professionals who are working in or have just completed studies in the foreign country concerned.

The problem of the migration of talent from less to more developed countries has attracted considerable attention. There have been many studies (both official and non-official), seminars, conferences and much has been published.

Dividends Ahead

It is gratifying, therefore, that all this energy and effort should at last have begun to show dividends. The statistical and conceptual uncertainties of the problem have been so narrowed that an authoritative report can present as its central finding the fact that: "... highly trained personnel from many developing countries are emigrating to a few major developed countries, that the size of this flow is large and that it is increasing at a rapid rate:"[4] It has therefore become less and less possible for individuals, institutions or governments to postpone the day when they will have to face up to this problem, on the excuse that the nature and extent of the migration and its consequences for the countries of emigration are not "adequately" known. There has also been increasing realization that this problem has implications for the developing countries which are just as serious as food shortages or overpopulation. In the past few years therefore several remedial measures of a non-discriminatory and non-coercive nature have been identified and formulated. And yet there is no action. Perhaps the developing countries have not committed "adequate" resources to the measures which are primarily their responsibility. But the fact that the rich nations have shown little recognition of their responsibilities cannot be denied either. Such a state of affairs is perhaps the best indication of the need for international action.

References

1. Dandekar, V. M., in *The Brain Drain* (ed. by Adams, W.), Macmillan, New York, 1968.
2. *Annual Reports* for 1965, 1966, 1967, 1968 and 1969 (University Grants Commission, New Delhi, 1966, 1967, 1968, 1969 and 1970).
3. *Indians Holding Faculty Positions in USA, Technical Manpower*, Bulletin of the Division of Scientific and Technical Personnel **10**, No. 7 (CSIR, 1968).
4. *Outflow of Trained Personnel from Developing Countries*, Report of the Secretary-General to the Twenty-Third Session of the UN General Assembly, November 1968, p. 55 (United Nations, 1968).

Adapting Technology—The Korean Case[*]

DR. HYUNG-SUP CHOI[†]

ACCORDING to Graham Jones,[1] the capacity of a society to assimilate new technology depends not only on adaptation of technology to the country's conditions, but also on the country's adaptation to the needs of the technology. Capacity to do so depends on sufficient awareness of those who represent that society, as pointed out by R. A. Solo.[2] I will devote myself in this paper to the problem of adapting a society to the need for development of adaptive technologies, but hope to be able to encompass in the discussion also adaptation of technologies to the requirements of society.

I will take the liberty of citing a number of Korean examples, so as to offer specific data before embarking on generalizations. The developments I have chosen are, to some extent, unique in their inception, implementation, and impact upon the nation. I do not intend, therefore, to propose them as a common pattern to be followed by all developing countries, although I believe that their relevance may well transcend socio-political and cultural differences among them.

I will try to present, in retrospect, some of the most important steps Korea has taken to adapt itself to the need to develop adaptive technologies, and to outline the course we have mapped out for the future.

Let us first examine the reasons justifying great efforts to make a society capable of developing adaptive technologies. Obviously, there are many such reasons, but I shall mention only two. First, there is the necessity of applying modern science and technology to a country's problems to improve the productivity in economic activities and the efficiency of its services; second, there is the need to create national capacity to cope with newly emerging problems, so as to free a society from the vicious cycle of underdevelopment. In so doing it is often necessary to overcome cultural obstacles, stemming from many sources, traditional and modern, indigenous and exogenous.

In this regard, there is no fundamental difference between societies and nations, particularly among developing ones. As stated by such economists as Schumpeter

* Based on an address given at the Technology and Development Institute Seminar on "Generation and Diffusion of Adaptive Technologies in Developing Countries", October 4–6, 1972, Honolulu, Hawaii.

† Minister of Science and Technology, Republic of Korea.

[1] *The Role of Science and Technology in Developing Countries*, 1971.

[2] "The capacity to assimilate an advanced technology," *American Economic Review* (1966).

(1949), Kindelberger (1965), Dewhurst (1965), and Quinn (1970), technology is not just one aspect of growth, but a vital factor in economic development.

Continued development of industry directed at creation of a self-reliant economy is possible only when supported by steady progress of science and technology. The designation "underdeveloped country" indicates technological backwardness as well as economic underdevelopment. Of some 130 countries in the world now, about thirty advanced countries, though they account for less than one-third of the world's population, dominate the world's science and technology. These advanced nations contribute the overwhelming majority of the world's research investment, of the world's scientific publications, inventions, patents and new products, and train most of the world's young scientists. Developing nations face the need to develop, simultaneously, their economy and their science. This is a difficult task, though ultimately rewarding. The underlying problems, especially in science development, are common to all developing nations, though in a varying degree.

The first problem common to all developing nations is the weak policy foundation for scientific and technological development and concomitant development strategies. In these countries, public interest in science and technology is low; their traditional cultures are hostile and pose obstacles to the establishment of a viable science policy. The growth of science requires an informed public, and popular esteem for the professions of scientist and technologist and for their accomplishments.

The second problem is lack of viable institutional setups, and inadequacy of research and development systems. This makes it extremely difficult, if not impossible, to induce talented people to stay with an institution and make it function efficiently. One of the last things to do in the initial development stage of a developing country is, it seems to me, to adopt blindly the "centers-of-excellence" concept, which may result in the spreading of talents, and fail to act as a breeder. I do not advocate, however, indiscriminate integration of research activities. Most developing countries, however, usually lack a central institution, or, if they do have one, it is either undersized or unbalanced and fails to reach a "critical mass". These countries have insufficient research equipment and low research budgets, and do not guarantee the autonomy of research establishments. A few existing research organizations are operated by governments, and their "research" includes testing and analyses. These countries are not able to use adequately and efficiently their limited number of scientists and their limited research funds.

The third problem is limited scientific manpower and, in particular, its skewed distribution, which makes it almost impossible to organize interdisciplinary research activities. Modern research demands a wide spectrum of specialized knowledge and experience, and unbalanced distribution of specialities can be a great hindrance to effective research programs.

The fourth problem is heavy reliance on imported technology. The technological development of a country naturally starts with the importation of advanced foreign technology, and proceeds through development of domestic variants of this imported technology to the final goal of technological self-reliance. After World

War II, Japan acquired many technological innovations from abroad. However, little can be expected from imported technology in absence of capability to modify and improve it for domestic applications.

The fifth problem is dearth of research equipment and technical information. The procurement of simple equipment pieces or parts can take months. The lack of technical information and absence of information specialists greatly retards research work. Hasty attempts at commercialization of research results often cause repetition of trials and errors, which lead not only to waste of money and energy, but also to poor relations between research and industry.

The sixth problem is poor capability for assessing the economic implications of research and development work. Economic feasibility studies are not performed before proposing research projects, and if performed, the findings are not always taken into adequate consideration. As a result, research projects often are not effectively tailored to their end purpose, and their results may have very little relevance to practical applications.

And lastly, but probably most importantly, there is often a lack of participation of relevant elements of society in the development of science and technology, particularly in the industrial sector, to which most of the applied research and development efforts are—and must be—directed. Without active involvement of industry, it is almost impossible to expect successful application of laboratory and bench-scale R & D to practical problems. This is, in my view, where the role of government should be enhanced, not only to make the pulse of industry felt within the R & D institutions, but also to make R & D a routine part of industrial activity.

This by no means exhausts the list of problems, the solutions of which are urgent prerequisites to successful development of science and technology in a developing nation. The solution to these problems must be given the highest priority in long range science policy.

What ought developing countries to be doing to equip themselves sufficiently to be able to get over the hump? I believe that it is essential for them to create infrastructures and environments conducive to the development of organized problem-solving capability, rather than indulging in random—and therefore ineffective—research efforts.

After 1945 liberation from Japanese occupation, Korea first suffered under severe economic and social ordeals; then followed the Korean War, which left behind utter devastation. During the 1950s our entire resources—what little of them had been left—and all of the foreign assistance had to be used for rehabilitation of the war-torn country and restoration of social order.

In the early 1960s, the Korean Government initiated an intensive development policy designed to industrialize the nation. Since industrialization can only be made successful by effective utilization of science and technology, the Government planned an intensive policy for the development of science and technology. Parallel with the Second Five Year Economic Development Plan (1966–71), we carried out the Second Five Year Plan for the Development of Science and Technology, with emphasis on manpower training, domestic technological development, and technical cooperation with foreign countries.

During the First Five Year Development Plan, the secondary industry sector—manufacturing and mining—had grown at a rate of 14.8% annually. This growth was further accelerated under the Second Five Year Plan, reaching an annual rate of 20.9%. Manufacturing industries had been the leading sector of the economy during the preceding decade, and were again chosen for a similar role in this decade. This implied the necessity for a similarly accelerated (and increasingly efficient) application of science and technology to the rapidly emerging industries.

Korea is not endowed with abundant natural resources. It has had a hiatus, so to speak, in recent history, years in which the country had no opportunity to develop its capabilities and to accumulate capital for growth; but Korea has a large population amenable to education and training, as a major asset for development.

The Korean Government decided to pursue a long-range plan for the development plan of science and technology. The Third Five Year Science and Technology Development Plan (1972–6) is an integral constituent of the Third Five Year Economic Development Plan (1972–6). In this plan, Korea adopted what might be called a three-pronged approach, emphasizing manpower development at various levels, accelerated introduction of foreign technologies, and stimulation of domestic R & D activities. We believe that domestic R & D is a prerequisite for sound selection, efficient assimilation, and fruitful adaptation of foreign technologies. This relation is widely recognized as valid for almost every developing country but is hard to use as basis for action. It implies that priority must be given to sound and efficient utilization of available foreign technologies, and that one must avoid engaging blindly in such noble but certainly financially unrewarding exercises as reinventing the wheel.

Korea enacted the Science Advancement Law and established the Ministry of Science and Technology to serve as central planning and coordinating agent of government in science and technology, to act as catalyst in the creation of the needed infrastructures and to help in devising measures to stimulate the development of science and technology for national development. In addition, the Government greatly increased its financial support to the Korea Scientific and Technological Information Center (KORSTIC), to make it responsible to the needs of Korean industrial and scientific communities. The Korea Institute of Science and Technology (KIST) was established in 1966 as a multidisciplinary industrial R & D institution.

In recognition of the need for an institution which would bridge the traditional and the contemporary, the domestic and the foreign, KIST was staffed with individuals who had had experience both in academic and in applied research and development activities, and equipped it with modern facilities.

We wanted to create a non-conventional, problem-oriented, self-perpetuating research group which would become the center of R & D activities, and act as a breeder institution, helping in the formation of scientific and technological communities throughout the country. Such an institute was particularly needed to work over some difficult-to-assimilate advanced technologies from abroad, and adapt them to local situations. This task requires on the one hand capacity to appreciate indigenous problems, and on the other capacity to find answers to these problems in exogenous sources.

In May 1965, during summit talks between President Park and President Johnson in Washington, D.C., they agreed on the establishment of a multidisciplinary industrial R & D institute in Korea, and President Johnson offered American support for it. In July 1965 a team headed by Dr. Donald F. Hornig, President Johnson's Special Advisor for Science and Technology, arrived in Korea to follow up the agreement.

The Governments of Korea and the United States of America engaged the services of the Battelle Memorial Institute to assist in the initial organization of the Korean Institute and in definition of the operations in which it should engage in its initial years. Subsequently, Battelle Memorial Institute was designated as a sister institution of KIST. Then, the Government of Korea and of the United States entered into an agreement which resulted in the organization of the Korean Institute of Science and Technology in February 1966. Along with the public announcement of KIST's establishment, President Park officially signed the Article of Incorporation, and I was appointed President of the new Institute.

So much for history; let us now examine the fundamental concepts used in the establishment of this new center for science and technology. After comparative study of various research institutes in advanced countries, it was decided that the new institute should be an independently incorporated contract research organization conceived in and infused with a spirit of responsibility. Consequently, autonomy, stability, and a research environment conducive to attracting needed manpower became key features.

Maximum autonomy in research is a paramount requirement for successful creative activities. In recognition of this fact, a special law, the KIST Assistance Act, was promulgated, according to which the Government was to donate money and land to KIST, but had no right to audit KIST accounts, nor to determine its annual programs. This law was amended in March 1967 to strengthen KIST's autonomy still further.

The ultimate authority in the Institute resides in a self-perpetuat board of trustees, composed of eleven members. Three of them represent the academic community, three the business community, and five are ex-officio members—the Vice Ministers of the Economic Planning Board, Commerce and Industry, and Science and Technology; a representative from the U.S. Agency for International Development in Korea (AID K), and a representative from the sister institution, the Battelle Memorial Institute.

Once autonomy is assured, stability can be sought. During its early, formative years, KIST was like an infant just starting to crawl, in need of strong governmental support until it could walk alone.

Up to March 1972 the Korean Government provided KIST with an endowment fund cf 2.3 billion won—equivalent to around 7 million dollars which was to be used for investment, with income from this investment helping to defray the operating costs of the Institute. Research expenditures, on the other hand, were to be derived from research contracts with government and industry.

In order to formulate the main thrusts of its initial endeavours, KIST had to review the industrial sector of Korean economy; on the basis of this survey, it was able to determine the personnel it needed to staff it, both in numbers and specialities. The survey involved an examination of the current state-of-the-art in application of technologies in Korean industry and identification of its needs. In this way KIST's role in the Korean setting would be placed in proper perspective. Over a period of 8 months, 25 different industrial sectors, involving some 600 industrial plants and related organizations were surveyed by some 80 engineers, scientists and economists, including 23 participants from Battelle.

The survey led to the following list of main operational areas for KIST in its initial years:

1. Materials and metallurgical engineering
2. Food technology
3. Chemistry and chemical engineering
4. Electronics
5. Mechanical engineering
6. Industrial economics

Attention had then to be turned to staffing to cover these areas. It was determined that there were quite a few Koreans well trained academically, but a severe shortage of those with industrial experience. From the outset, we were greatly concerned with the problem of how to induce the return home of Korean scientists and engineers residing abroad, despite wide discrepancies in salaries and differences in living conditions. We anticipated great difficulties in bringing about this repatriation.

Contrary to existing expectations, we received, however, favorable reactions from about 500 people out of a total of 800 to whom brochures and employment applications were sent. KIST and Battelle screened the list, and selected sixty-eight candidates. Of these, KIST selected twenty-five who had had at least 5 years of relevant experience beyond doctorate degrees or other equivalent, qualifying them for principal investigatorships at KIST. Of these twenty-five candidates, eighteen signed employment contracts. In addition, nine senior-level scientists were recruited who had completed their academic training abroad, but were employed in Korea. Thus we secured twenty-seven senior research staff members by the end of 1968 in what we called "counter brain-drain" operation. Twenty more senior-level staff members have been added to date to this initial group.

Also to be determined were physical facilities—buildings and research equipment—needed for the staff's work. From considerations of the staff requirements, the requisite buildings and research equipment, and the program of the Institute, we derive the fiscal requirements of KIST.

During the first 5 years of KIST, from its establishment in February 1966 through February 1971, emphasis was placed on expediting construction, establishing the management system, choosing research fields, securing research staff, and acquiring research equipment. Research activities did not go into full swing until the latter half of 1969, when construction of buildings and facilities was completed.

Being a contract research organization, it was necessary to adopt several management systems which governmental institutes do not often employ. These were, first, establishing a Project Development Department as the vehicle for selling the concept of scientific research for solving the problems of industry and as a means of assisting industry to formulate questions to be asked of the Institute; second, adopting a cost accounting system aimed at self-support of each R & D operation by gauging input as compared to performance including logging use-rates on equipment, space, and utilities (the Institute itself underwrites some basic research projects, which may not have industrial support, but are considered relevant to anticipated industrial applications); third, adopting a system of independent laboratories as basic organizational units. (A number of laboratories devoted to specialized fields were created in response to R & D projects already contracted for, or foreseen to be contracted for. Each laboratory is managed independently, under the direction of a principal investigator. Operation of each laboratory and its research procedures are based on cost accounting, including the costs of utilities use, prorated for all utilized equipment and facilities. Principal investigators manage their research expenses without restraint by any outsider, and assume responsibility for the results. Despite this autonomy, a flexible, coordinated, interdisciplinary plan is devised, determining the dominant thrust of KIST's activities). The fourth system adopted was a contract appointment system. A research staff member joins the Institute by signing a contract after evaluation of his research capacity and of the contract research theme. After the first 2 years, renewal of the contract depends on demonstrable value of the work performed. Depending on circumstances, commissioned outside investigators may be temporarily employed. Staff members are given salaries adequate for decent living, and provided with modern housing facilities, so that they may devote themselves unreservedly to their research.

In consideration of the importance of developing the capabilities of our research staff, the Institute is granting a 1-year sabbatical leave to principal investigators for each 3 years of continuous service, permitting them to familiarize themselves with overseas science and technology trends.

As compared to 1968, research contracts received by KIST in 1971 have increased 4.5 times in number, to 189 projects, and 3.5 times in value. For the first 8 months of 1972, research contracts rose to the equivalent of almost U.S. $3.4 million. If this trend is maintained, the 1972 total should be 16 times that of 1968.

An upward trend is shown in the average magnitude of research contracts, now amounting to about $23,000. The Government share remains about two-thirds of the total sum, and about one-half of the total number. Of the remaining one-half, small and medium industries claim about 60% in number and 66% in amount. Mechanical engineering leads in contract amount with 25% of the total, followed by electronic engineering, chemical engineering, engineering services, computing services, metallurgical and materials engineering, and food technology, in this order. Computing services lead by far in the number of projects, they are followed by electronic engineering, chemical engineering, metallurgical and materials engineering, and food technology.

A total of 53 out of the 209 contracts received in the field of manufacturing technologies have already gone through bench-scale testing, or will be tested in the near future. Among them are the manufacture of titanium dioxide from Korean ilmenite, single-cell protein from petroleum hydrocarbons, freon 12 from Korean fluorspar, ampicillin, printed circuit boards, and synthetic wigs. There are twenty-three cases where R & D developments at KIST are either ready for application to commercial production or are awaiting commercialization.

At present, KIST has drawn royalties for sixteen cases of commercial application resulting from its R & D, amounting to $40,000. These include ferrocement structures, and the regeneration of used lubricants from internal combustion engines.

Now, I will attempt to make a qualitative assessment of a few selected projects in which KIST has played an important part. I will break them down into four categories: engineering services, R & D for industry, R & D for government, and R & D for foreign agencies. I will cite specific examples in the category of general engineering services for industry.

First, I cite an engineering service contract with the Pohang Iron and Steel Company, which is building Korea's first integrated iron and steel mill with an initial production capacity of 1.03 million tons of crude steel per annum. KIST has engaged the services of some foreign specialists, mostly from Japan, to supplement its own staff, and provide necessary technical services in the construction and operation of this plant, including specifications for facilities and equipment. It played a central role in the formulation of detailed plans for this steel mill. As a result of this service contract, the company has saved a considerable amount of foreign exchange, which it would have had to spend if it had engaged a foreign consultant firm.

A second example is that of helping an industrial plant to adapt foreign technology. The Korea Special Metal Company, which produces bearings for automobile engines, engaged KIST services to assimilate and adapt the technical know-how to be provided by a foreign firm. Since introduction of technical know-how from abroad lagged behind schedule, the plant had run into many technical problems, ranging from the sintering of copper–lead alloy powder to electroplating of metals.

Through a combination of in-plant diagnosis of technical problems, and in-plant instruction of workers involved in application of the new methods, a number of improvements were made based on KIST laboratory testing and experiments; for example, the optimal sintering atmosphere for copper–lead alloy powders was determined and better gravity casting procedures put into effect, resulting in improvement of the cast bearings. A new application of tri-element electroplating techniques to the fabrication of bearings resulted in improvement in their performance.

Third is a case in which KIST was called upon to render crash engineering service to a polyester spinning mill, suffering from malfunctioning of unknown origin. Immediate help was asked for from an engineering firm in Europe by the Korean company, but the response was discouraged in that they could provide such service only after a couple of months and for a fee of $200,000—almost one-third of the total cost of the installed equipment.

KIST was then called in. Having identified the factors responsible for quality of the spinning, the control systems were modified to ensure their proper functioning and optimal performance. This, along with increased spinning speed, raised the mill's production capacity from 1 ton to 1.5 tons daily, while lowering the discards from 40% to 3–5%, at the cost of less than $30,000, payable in local currency.

Of the many research projects contracted with KIST by private firms, I shall cite three examples. First, development of a remote control tuner device for T.V. With a contract amounting to a little over U.S. $15,000, KIST developed a tuner containing locally produced VVC diodes to replace the conventional rotary switch, and also introduced a numeric memory remote-control device. The VVC tuner makes instant channel changes and eliminates the breakdowns common to conventional tuners.

Second, I refer to several successful syntheses of organic pharmaceuticals, which have led to the construction of commercial-scale plants, permitting not only substitution for imported compounds, but also marked reduction in costs, sometimes by as much as a factor of 100.

Programs undertaken by KIST at the request of the Government include, among others, a program for the advancement of electronics industry. KIST was designated as one of three organizations charged with responsibility for fostering the development of Korea's electronics industry, particularly through R & D activities and information analysis. In this connection, KIST has already completed sixty-six projects, which comprise almost half of its total R & D activities in the electronics field.

One noteworthy program KIST has undertaken in collaboration with a foreign government is NASA Technology Transfer Pilot Program, which has continued for the last 2 years. Although in the nature of a pilot project, the program encompassed a whole spectrum of activities, ranging from identification of problems, through search for appropriate technologies to solve those problems, to research, engineering, manufacture of prototype products, and commercial application. The experience acquired in the course of this highly stimulating work was presented at the International Seminar on Technology Transfer held in November 1972 at KIST.

So far, I have been enumerating some of the most satisfactory KIST programs, either contributing directly to the development of adaptive technologies, or providing important anciliary "spill-over" effects, such as helping to form a viable community of scientists and engineers, and establishing a social milieu favorable for the development of science and technology.

Despite detailed preparatory work, almost unprecedented in its completeness, to bring KIST into operation in the context of economic development of Korea, it has also experienced many trials and made some errors, which can be avoided in the future.

While KIST was primarily brought into being to help industry, and despite its efforts to achieve this purpose, the interaction between KIST and industry, while encouraging, needs to be extended at a faster pace. In this respect, efforts of KIST alone have a limit; in this area, the Government should play an increasing role, through legislation and by underwriting joint programs. With an organization like KIST to help it, government could lead in analyzing the technological aspects of many of its development policies, with which Korean society as a whole has had no past experience.

KIST, in my belief, has fulfilled original expectations, particularly by its spill-over effects. In tangible contributions to development and adaptation of industrial technologies, KIST's success cases are numerous in consideration of its short history, but many more of them are required for a significant impact upon Korean industry as a whole.

As had been hoped for by those who took part in the inception of the Institute, the concept of a multidisciplinary R & D institution, chartered and operated in the way KIST is, enjoying support by the legislature and by the President of the nation, proved to be a viable proposition. It may become essential for other developing countries as well, if they plan to use science and technology as important factors in their economic growth.

Now we come to other Korean Government activities essential for science and technology to take their proper place in national development.

The Ministry of Science and Technology has been engaged in strengthening the foundations of science and technology in Korea and creating an environment in which people engaged in scientific pursuits and skills will be respected. To provide impetus for the development of science and technology, the Ministry formulated a Third Five Year Manpower Development Plan, emphasizing the need to develop technical manpower at all levels.

To meet the pressing need for high-caliber manpower in science and technology, the Korean Advanced Institute of Science (KAIS) has been established, at Government's expense but as an autonomous entity, and is a neighboring institution to KIST, in the hope that the two institutions will collaborate to the maximum feasible extent.

In addition, the Korea Development Institute (an economics research institute), the Agency for Defense Development, and the Korean Scientific and Technological Information Center, have all been located in the same area, forming a "Science Park".

To ensure cooperative and collaborative policies among these institutes, the Ministry initiated the formation of a Consultative Committee, consisting of heads of the Institutes in the Park, to discuss matters of common interest and to effect the sharing of both physical and intellectual resources. Using KIST as the nucleus, a community of intellectual, scientific and technological people and facilities, specifically oriented toward the problems of Korea, has been assembled. These organizations are chartered and operated to absorb liberally developments from advanced countries, but also to create within the Korean social and physical environment bases for the building up of internal capabilities that can be used for future economic advantage, both in Korea and abroad.

One of the most important items of legislation presently sought is a Law for the Promotion of Industrial Technology Development. Its objectives are twofold. First, to induce industries of all sizes to undertake R & D developments by providing fiscal and financial incentives, and to stimulate mutual help through sharing of expenses. This involvement of industry can play a role at all stages, from project formulation to commercial application of R & D results. Secondly, the law is designed to facilitate introduction of foreign technologies into Korea by minimizing the administrative procedures involved, and to obligate the end-user of imported technology to adapt it to Korean conditions, either by himself, if he is adequately equipped with staff and facilities, or with the assistance of a government-designate organization such as KIST, or another adequately equipped R & D institute.

In suggesting such provisions, we have taken note of experiences of other countries. The United States of America, in the late nineteenth century, had to import many technologies from Europe to provide a basis for its industrialization. Japan, for the last decade or so, has been spending the equivalent of 2% or more of its total export income for the importation of foreign technology. To me, it is no coincidence that close correspondence exists between the level of foreign technology import into Japan, and the level of commodity exports out of that country.

I am further impressed by the fact that Japan has been spending as much or more on adaptation activities within the country as on technology importation. I am of the opinion that without accumulated resources of science and technology within the developing country, and without adequate research and development efforts, there is a limit to importation of foreign technology, however good it may be. Innovative R & D activities have to be maintained to preserve the viability of an industry. This holds particularly true in the case of developing countries, where it is necessary to depend to a large extent on foreign technology.

In closing, I would like to underscore the statement of Stevan Dedijer[3] that positive involvement of a chief executive of a nation in the development of science and technology for economic growth and social development, is not merely desirable, but essential for developing countries, if they are to develop their science and technology and apply them efficiently to their development. Korea has been most fortunate in the understanding of, and support for, science and technology, by its Chief Executive.

Thank you.

[3] *Underdeveloped Science in Underdeveloped Countries* (Minerva, 1963).

*The Scientific Community and Latin American Development**

JOSE BARZELLATO, M.D.†

THE central idea under which this Symposium has been convened can be summarized in Pope Paul's famous sentence: "Development is the new name of peace."[1] Some speak simply of "development", others of "integral development", or of "human development", but gone are the days when the only preoccupation was "economic development". It is not necessary any more, especially at a meeting like this, to emphasize that economic development must include social justice, nor that science and technology are an important component of true development. Let us only say that the key problem of development is not to create wealth, but rather to build a capacity to create wealth, provided the results benefit all the people of the country concerned.

I would like to start by saying that nowhere in the world is underdevelopment such a threat to world peace as in Latin America. Precisely because our nations are not the most underdeveloped countries in the world, they are experiencing the most violent "awakening of expectation". "Our people know that they live poorly in a rich continent . . . We not only have illiterates but also the aspiration for knowledge and the awareness that the present civilization has created the possibility of acquiring that knowledge." I am quoting from the speech of one of the Latin American Presidents at the Meeting of American Chiefs of State (at Punta del Este, Uruguay) in 1967. He also said: "Something of very deep meaning is at stake in our America: to know if within the coming years (and I emphasize the word 'coming', as opposed to the next century), those who love liberty and believe in the dignity and essential rights of the human being, those who think that man is an end and not an instrument, will be able to organize, simultaneously, both economic and social development and thus give an answer to the just expectations of so many people." Furthermore, this president, President Frei of Chile, added: "I am convinced that for most of our nations, if not for all of them, it would be difficult—I do not dare say impossible, although I think so—to achieve such an answer if they stay isolated, no matter what the extent of their wealth and size."[2] This is the challenge to which the scientific community can contribute; but, before making some suggestions in this respect,

* Presented at the Pugwash Symposium on Science and Development, Stanford University, September 1970.

† Special Adviser to the Chairman, CEPCEIC, Organization of American States, Washington, D.C.

allow me to briefly sketch what were, until recently, the principal characteristics of scientific and technological development in Latin America.

The scientific and technological infrastructure in Latin America is weak, particularly in the quantity and quality of its human resources. Hence it is obvious that remedying this situation is of the first priority. Nevertheless, other things must be done simultaneously. Traditional Latin American cultural values have not assigned the same importance to scientific and technical creation as to other aspects of human creativity, generating an atmosphere of indifference to science and technology, in which only a very small percentage of graduates of the institutions of higher learning have decided to dedicate their lives to research. Real scientists have been few, usually poorly paid and socially under-esteemed, and practically always dedicated to basic research. They have in general been isolated from, and even indifferent to, the concrete needs of the society in which they live. Technological research, the natural link between basic science and the needs of a society, has been almost entirely missing.

The amount of research done has been limited, and has been almost exclusively carried out in public institutions, primarily in the universities. These have almost no contact with government and its plans, or with the private sector of the economy and its needs. Scientists looked for their reward in the pursuit of knowledge, and in the recognition of their foreign peers, while the university saw in them only the asset of prestige. Industry—mainly public in many countries, but whether public or private, usually monopolistic and developed under a policy based on substitution for imports supported by strong protectionism—lacked the competitive climate that helps to create the demand for innovation. Such an origin also explains the small volume of most industries, in most cases created to satisfy a very small market, thereby imposing financial limitations on research possibilities. The practical long-range consequences of this situation are reflected, together with other short-run causes, in the present crisis of Latin-American exports. These are experiencing a steady decline in percentage of the volume of world trade. Furthermore, governments have not created incentives to stimulate native innovation.

Under such circumstances it is easy to understand why the importation of technologies is such an important process, even though it is usually carried out through enterprise agreements which are, in the case of small enterprises, frequently poorly selected, generally bought at too dear a price, and seldom adapted to the local characteristics. In addition, these agreements rarely have an impact outside the particular enterprise involved, due to the low capacity for the diffusion of innovation (the so-called "technical dualism").

This has been, in general, the picture of the development of science and technology in Latin America until recent years, with the exception of only two sectors: medicine and, in some countries, agriculture. Only in these fields, and especially in the area of public health, have there been national policies integrated into national development programs, reasonable scientific and technological infrastructure and adequate, rapid transfer of technology, with proper adaptation and diffusion. It is interesting to note that in medicine—with the exception of the pharmaceutical industry, which is not directly involved in the organization of health care—there is no application of the patent system.

During this last decade, and with progressive intensification, this situation has clearly been changing. All university forums, at both the student and faculty level, are now speaking about the responsibility and role of the Latin American university in society. The number of university full-time personnel is steadily increasing. The main growth is oriented toward science and technology. Technological Institutes and government research institutions are being created, some of them oriented towards crucial aspects of development.

At the political level there has also been a spectacular change. I have already mentioned the Meeting of American Chiefs of State in Punta del Este, Uruguay, in 1967, which marked the beginning of a regional concern for the development of science and technology. The Presidents proclaimed their decision to "harness science and technology for the service of the people of the Hemisphere".[(3)] The Action Program they approved included a special chapter in which Inter-American co-operation was called for to complement national efforts in this area. Specifically, the idea of a Regional Scientific and Technological Development Program was approved. It is encouraging to see that, 3 years later, such a Program is already in existence and has spent $6.15 millions in its first 15 months of operation, having already sent 213 visiting professors and researchers to Latin America and 319 Latin American fellows and research assistants to the institutions chosen as regional training centers. Its budget for the financial year 1970 is to reach 15 million dollars.

This attitude at the political level was not a passing one. In May 1969 Latin America spoke to the new administration in the United States Government through the "Consensus of Vina del Mar".[(4)] This document includes 9 pages of proposals, 3 of which refer to scientific and technological development. The specific requests stated there are being discussed, and some of them are being implemented within the Inter-American System. Furthermore, at the First Meeting of the OAS General Assembly in 1970, the need for, and importance of, scientific and technological development in Latin America was clearly emphasized by the Minister of Foreign Relations of Argentina.

This new attitude in Latin American politics is also apparent at the national level. Eleven of the countries south of the Rio Grande, with the strong participation of the scientific community, have created governmental bodies to deal with policy-making for, and promotion of, science and technology. Most of them have been recently organized, but in a few countries they have been active for some years and already have significant budgets and functions.

Despite all these signs of change, many of the negative aspects previously analyzed are still present in varying degrees in most countries in Latin America, and, I suppose, in other underdeveloped countries; and we should keep them in mind when discussing the possible contributions of the international scientific community, and the Pugwash Conference in particular, to development.

Such a contribution, as in any other type of international cooperation, will only be successful if it is a complement to the efforts being carried out by the underdeveloped countries themselves. In other words, the developing countries must make a political decision—each according to its own model—to orient their effort towards development, and the developed nations should generously complement this effort as they are requested to do so. Mistakes and wrong attitudes on the part of the small countries can best be corrected through a free exchange of ideas; but when aid is tied to conditions which reflect the preconceived notions of the developed countries, two results may occur. One is that the objective may be reached only in part or not at all. More certain is that the countries will resent the procedure.

It is at this creative level of exchange of ideas and of finding solutions where I believe the greatest contribution of the international scientific community can be made. There are no accepted prescriptions to accelerate development, and I am quite sure that magical universal solutions will never be found, simply because each developing country has its own changing problems and characteristics, quite different from the rest. There is a great need for good research and analysis and, especially, for new ideas. No other forum could be less biased and less politically suspect in providing an arena for the identification and analysis of the problems that slow down development, and in suggesting solutions, than one made up of members of the scientific community.

The Pugwash Conference could organize a forum which could help to mobilize the scientific community of the underdeveloped countries and aid them in making their voices heard by their own governments. At the same time it would provide an opportunity to impress upon the governments of the developed countries a message containing such ideas as those I recently heard expressed by Albert Sabin, and I quote: "I would like to submit that what is on trial today is not science, but the leaders of the great powers who are still pursuing outmoded concepts of international relations that prevent the proper utilization of the fruits of science and technology for the benefit of *all* mankind." "It seems to me that an entirely new philosophy of international relations and aid to impoverished nations must be developed by both the capitalist and communist 'have' nations of the world. The capitalist and communist 'have' nations alike must come to realize that they have a common enemy in the poverty, hunger and despair of more than two-thirds of the world's population."[5] Just as developed countries are awakening to the need to pay the costs of maintaining an environment in which they can live healthily, in their relations with the developing countries they should realize that there are costs involved in living with their underdeveloped neighbors in a world of peace and understanding. And, finally, that it would be well to pay these costs before they become costs in human lives, as in Vietnam.

The need for a forum for relating science and technology to development is the result of a communications gap which must be overcome. First, among scientists themselves, to establish the meaning of their work and a kind of self-orientation and control for constructive purposes. Second, between scientists and the rest of the world, especially the developing world, to the end that science can make sense and be adequately controlled through political processes. I believe the warning made by Hannah Arendt is still a valid challenge to the scientific community. She wrote: "The reason why it may be wise to distrust the political judgment of scientists *qua* scientists is . . . precisely the fact that they move in a world where speech has lost its power . . ." that there is "a crisis within the natural sciences themselves. The trouble concerns the fact that the 'truths' of the modern scientific world view, though they can be demonstrated in mathematical formulas and proved technologically, will no longer lend themselves to normal expression and thought . . ." and that "men in so far as they live and move and act in this world, can experience meaningfulness only because they can talk with and make sense to each other and to themselves."[6] What more elegant justification than this is needed for conferences such as Pugwash, or for the extension or elaboration of the Pugwash concept in relation to science and development?

References

1. Pope Paul VI, *On the Development of Peoples* (*Populorum Progressio*), p. 51. Washington, D.C.: U.S. Catholic Conference, 1967.
2. Organization of American States, Pan American Union. *Final Report of the American Chiefs of State, Punta de Este, Uruguay, April* 12 *to* 14, 1967, "Remarks of the President of Chile, Eduardo Frei Montalva, in the Inaugural Session (First Public Session) celebrated April 12, 1967" (OAS/Ser. C/IX.1), pp. 141–145.
3. *Ibid.*, "Declaration of the American Presidents", p. 4.
4. Organization of American States, Inter-American Economic and Social Council. Special Meeting of CECLA at the Ministerial Level, Vina del Mar, Chile, May 15 to 17, 1969 (CIES/1403).
5. Albert Sabin, Inaugural address to the Israel-Latin American Symposium on Science and Technology and Organization of Research, Jerusalem, January 1970.
6. Hanna Arendt, *The Human Condition*, pp. 3–4. Garden City, New York: Doubleday & Co. 1959.

*Some Obstacles to the Effective Utilization of Trained Scientific and Technical Personnel in Developing Countries**

L. K. H. GOMA†

Introduction

The experience of developed and of some developing countries has shown that development is possible largely through the application of science and technology; and this is obviously affected by the supply and maintenance of trained manpower of high quality. The acute shortage of scientific and technological manpower in the developing countries is, therefore, perhaps the most serious handicap facing some of these countries in their development efforts. What is highly doubtful and disturbing, however, is whether effective utilization is made, by these countries, of the limited scientific and technological personnel available to them. Thus, in the case of Africa for instance, the available data would seem to indicate that there is under-utilization of nationals with scientific and technical training.

It would not be altogether correct to say that the entire trained scientific and technological manpower (however scarce) available to the developing countries has not made significant contributions to the task of national development. However, there are formidable obstacles which militate against the utilization of such scarce human resources to maximum advantage. To overcome the obstacles it is important first to recognize and examine these. This is the purpose of the present paper.

The Obstacles

Some of the obstacles, such as the "brain-drain", the isolation of the scientist, the shortage of competent technicians, and inadequate financial support, have been the subject of considerable discussion at international conferences and in the literature; others appear to have either been stressed insufficiently or received no attention at all. These fall under four main groups: (1) the general shortage of high-level manpower itself; (2) the unripe, uncongenial and sometimes risky environment; (3) academic freedom and university autonomy; and (4) the divided world and the international bandwagon.

1. THE GENERAL SHORTAGE OF HIGH-LEVEL MANPOWER ITSELF

Among the administrators in the developing countries, many have to make decisions which deal with scientific and technological matters such as health projects,

* Paper presented at the eighteenth Pugwash Conference on Science and World Affairs, Nice, France 1968.

† Vice-Chancellor, University of Zambia, Lusaka, Zambia.

new industrial processes, improved agricultural methods, development of power resources and many others (Moravcsik, 1964). An insignificant number of these decision-makers, if any at all, are sufficiently well versed in science and technology to be capable of making sound, realistic and prompt decisions. And in the absence of suitable advisory committees (itself due to the shortage of scientific and technological manpower) upon which the administrators can rely for expert advice, the few available working scientists and technologists may find themselves undertaking improperly planned and pointless programmes that can only undermine their morale. The right choice of scientific/technological problems and the successful prosecution of the problems chosen are major influences on the attitude of scientists/technologists toward, and their effectiveness in, their work (Kapitza, 1966).

Sometimes, and because there is no one else, scientists and technologists are channelled into employments not commensurate with their training, qualifications and merit. Similarly, in many cases where the social and economic structure has not been adjusted in time, large numbers of young people trained in science and technology can at first find employment only in posts below their capacities.

The few available trained scientific and technical personnel have to grapple with too many routine duties and too large a range of problems requiring immediate attention, with the consequent frittering away of the limited efforts and resources. In some countries, organizational bureaucracy has resulted in too many research institutes; consequently, the few scientists and technologists find themselves becoming directors of these institutes and thereby removed from doing active research themselves.

2. THE UNRIPE, UNCONGENIAL AND SOMETIMES RISKY ENVIRONMENT

The available manpower needs some definite atmosphere in which to work and in which it can produce of its best (Wasawo, 1965). However, it is clear that in the vast majority of the developing countries, there has generally been too little experience of scientific and technological work to permit a realistic image of what is compatible with their effectiveness (Dedijer, 1963). As a consequence, the available personnel often find themselves working in an atmosphere charged with prejudice—which may frustrate the prospects for individual promotion or the possibility of study leave elsewhere for workers of recognized calibre; the distrustfulness of politicians; overt bureaucracy and antipathy to the needs of scientists for freedom from petty controls. These restrain scientific zeal and productive interest.

Research, of whatever kind, cannot thrive where there is no sustained demand for it. Ignorance of what is attainable through science and technology, general apathy, and the consequent lack of or insufficient demand for research by industry, agriculture and other sectors of the economy in the developing countries thus militate against the effective use of the limited human resources.

In many developing countries foreign pressures and domestic centrifugal forces give rise to grave political, economic and social strains and instabilities (Dedijer,

1963). This may further lead to loss of earnings because of prohibitive taxation or, for foreign personnel, refusal to allow sufficient or any repatriation of funds; while racial or tribal conflict may result in administrative collapse (H.H. The Aga Khan, 1966). Drastic and severe reprisals against the universities and the persecution by military governments of university professors because of their ideological stand may occur frequently in some countries (Pozo-Olano, 1967). All these "high risks" are a constraint further complicating the conditions under which the few available scientists and technologists must work and which must, therefore, cause much damage to the general progress of science and technology in these countries.

One of the striking characteristics in the developing countries is that the personnel on whom the science and technology depend, as far as modern research is concerned, consists mostly of quite young persons (Moravcsik, 1966). The lack or inadequacy of proven and "ideas men" to stimulate and guide the young scientists and technologists must inhibit their effectiveness, as considerable waste of time and effort could and does result before young people acquire enough experience to carry out independent research themselves.

Although the importance of a central organization to determine priorities and co-ordinate the scientific and technological efforts of a country has been accepted by the developing countries, the sad fact is that many of them still do not have such a body. The effect of this, as implied above, is that the limited human resources are often wasted on meaningless programmes—programmes that lead nowhere: they neither contribute to the advancement of science and technology nor do they have any relevance to meeting the urgent needs of the country. Moreover, duplication of effort is possible, under these conditions, since it is not easy to know what is going on elsewhere or to relate one's work to the overall needs of the country.

Insufficient decolonization of the mind and the consequent uncritical adulation of foreign experts and know-how presents one of the biggest obstacles to the effective utilization of national or local scientists and technologists in many developing countries. The work of nationals is often not seriously appreciated or demanded by government, industry, agriculture or other sectors of society. Instead, specialists are brought in, at very great expense, to produce, in a few days, reports that are often worth next-to-nothing (Gouin, 1966). In some cases, the advice that the visiting dignitaries give today may be identical with what the national expert has been saying for years, but which has never been listened to (Mukerjee, 1964). The doors which readily open to the foreign experts are all but shut to their domestic counterparts. Equally relevant is the fact that where industrialists (who may themselves be non-nationals) do use advanced technology, they tend to depend on what has been or can be imported from abroad, often for prestige, and the expected rapid economic returns (Nayudamma, 1967). This lack of self-confidence or self-esteem in their country is most damaging to the morale and effectiveness of national scientists and technologists.

A related problem is that of the area of recruitment. Under colonial rule recruitment for the developing countries was largely from the colonizing metropolitan countries. Those trained or coming from elsewhere were deemed to have received an inferior education and, consequently, were denied employment. Regrettably, such an attitude continues even after independence in many of the developing countries.

3. ACADEMIC FREEDOM AND UNIVERSITY AUTONOMY

It is now widely accepted that universities in the developing countries should be important instruments of general development. Not only must they produce the high-level manpower required by their countries, but also, since in most of these countries the universities are staffed with a large proportion of the very few experienced scientists and technologists available, utilize their personnel and the physical plant for the maximum benefit of the country. However, in general, the latter is proving extremely difficult to accomplish; and it would appear that the concepts of academic freedom and university autonomy are among the important bottlenecks here. Although many of these universities have expressed a desire to be responsive to the real needs of their host countries, few have made serious attempts to become acquainted with national aspirations, resulting in lip-service responsiveness, which may itself give deceptive satisfaction all-round.

In most of the developing countries the universities are largely dependent on a university community outside their borders (Pieris, 1964; Silcock, 1961, 1964). As a consequence, the university personnel, in many respects, find themselves hampered from being positively responsive to the real needs of the country by academic snobbery and feelings of professional insecurity: for many of them individually, so much depends on the impression they make abroad; attention to or involvement with a local research tradition earns no credit for personal future advancement elsewhere. Many can only base their claim to "superiority" on a Master's, or Ph.D., degree obtained abroad.

The maintenance of high academic standards or their attainment, where these have yet to be established, must be guaranteed. However, and regrettably, preoccupation with the observance of international standards (and hence, concern with the university's international standing) has often resulted in many universities in the developing countries "thriving" on anachronistic and unimaginative practices. Standards, as Silcock (1961) points out, are made up of many components; and these are affected in different degrees and different ways by differences in situation. This does not seem to be fully appreciated; to many academics in these universities "the idea of a standard as something measured in one dimension, capable of being tested by simple comparison, has become an academic straight-jacket, hampering . . . meaningful research".

In certain cases (Haniotis, 1968) not only has the senior level of the academic profession adopted a seemingly deliberate policy of isolation from efforts towards national economic development; but has also been energetic in retarding efforts to reform higher education. In this it has been sustained by the constitutional provision of university autonomy originally developed in quite different environments. For some indeed autonomy has been misused as a cover for indolence.

The desire for relevant scientific and technological research often evokes the problem of basic versus applied research in relation to where it should be done. There are many in the universities who regard applied research as falling outside the range of proper university endeavour. And yet, what seems to some of us to be an outstanding attraction of the developing countries for good research-minded academics is the abundance of worthwhile problems—problems that should be the subject of meaningful research, highly challenging and rewarding academically and at the same time directly relevant to the development efforts of the country.

4. THE DIVIDED WORLD AND THE INTERNATIONAL BANDWAGON

It has been said, with considerable justification, that international co-operation in science and technology, so far, has been politically important but scientifically negligible. High-sounding resolutions have been made, but their implementation has been frustrated by the hostile tendencies of the divided world. For the developing countries this is extremely disappointing.

Competing programmes and aid as a tool to align nations in a divided world have frustrated the real benefits which the policy of genuine non-alignment could bring to the developing countries in the field of science and technology. In a bid to get a footing in a developing country, certain developed nations have been only too eager to jump on the international bandwagon without any serious consideration of the real needs of the developing country. This often results in the provision of incompetent and unimaginative "experts" by the aid-extending countries. Similarly, wrong and conflicting advice is given; and the developing country may thus embark on programmes contrary to its interests, but which serve the interests of the donor countries.

International scientific and technological programmes can be of great significance in the development efforts of the developing countries. However, these programmes are insufficiently co-ordinated and may even set severe problems of choice for these countries. Important as some of the international projects are for the general scientific and technological advancement of the world, it is wrong if considerable amounts of the limited budget (earmarked for science and technology) and human resources available to the developing countries are taken up with international schemes which, as far as can be seen, are immediately unhelpful, restrictive and unrealistic, in relation to the general development plans of these countries. And there are a number of such schemes.

Paradoxically, instead of providing encouragement and a stimulus to the morale of the local scientists and technologists, and mitigating the frustrations of their isolation, visits from scientists and technologists from abroad often have the opposite effect. They can be too frequent, pointless and, hence, time-wasting. Many visitors do not seem to be briefed adequately on the particular country's problems; while the local scientists/technologists are not informed in good time of the impending visit, with the result that the "dropping-in" serves no useful purpose to either.

Concluding Remarks

The developing countries cannot afford to waste the limited scientific and technological manpower that is and can be made available to them. It is equally clear that the rate at which the obstacles to the effective utilization of this manpower can be overcome will vary from one country to another. However, the available manpower needs some definite atmosphere in which to work and in which it can produce of its best. A considerable number of the obstacles listed are largely psychological and a result of intellectual bigotry—they must be tackled through educating the mind; others have various origins as indicated above, and, to overcome them, requires the existence of a scientific community, proper coordination, an increased awareness of what is attainable through the application of science and technology, and a better appreciation of the real needs of developing countries and a better sense of proportion at the international level.

References

BUSIA, K. A. (1962) *The Challenge of Africa*, chap. 4, pp. 51–63. New York: Praeger, Inc.

DEDIJER, STEVAN (1963) Underdeveloped science in under-developed countries. *Minerva*, **II** (1), 61–81.

GOUIN, P. (1966) Co-ordination of scientific research in developing countries. *Proc. 15th Pugwash Conference*, pp. 178–85.

H.H. THE AGA KHAN (1966) *Education for Development—Nairobi*. East African Institute, 16 pp.

HALSEY, A. H. (1963) Science and government in Sweden. *Minerva*, **II** (1), 54–60.

HANIOTIS, GEORGE (1968) The situation of the universities in Greece. *Minerva*, **VI** (2), 163–84.

KAPITZA, PETER L. (1966) Scientific policy in the U.S.S.R. *Minerva*, **IV** (4), 555–60.

MORAVCSIK, MICHAEL J. (1964) Technical assistance and fundamental research in underdeveloped countries. *Minerva*, **II** (2), 197–209.

MORAVCSIK, MICHAEL J. (1966) Some practical suggestions for the improvement of science in developing countries. *Minerva*, **IV** (3), 381–90.

MUKERJEE, D. (1964) Indian scientific policy. *Minerva*, **II** (3), 360–8.

NAYUDAMMA, Y. (1967) Promoting the industrial application of research in an underdeveloped country. *Minerva*, **V** (3) 323–39.

PIERIS, RALPH (1964) Universities, politics and public opinion in Ceylon. *Minerva*, **II** (4), 435–54.

POZO-OLANO, J. (1967) The role of the university in Latin America. *Pugwash Newsletter*, **5** (1), 30–32.

QUIRINO-LANHOUNMEY, J. K. (1967) The great dilemma of international co-operation. *Pugwash Newsletter*, **5** (1), 32–33.

SILCOCK, T. H. (1961) Academic standards and the new universities of South-East Asia. *Science and Freedom*, **19**, 7–14.

SILCOCK, T. H. (1964) The development of universities in South-East Asia to 1960. *Minerva*, **II** (2), 169–96.

WASAWO, D. P. S. (1965) The nation and the university. *E. Afr. J.*, pp. 4–5 (Aug. 1965).

*Underdevelopment, Science and Technology: the Point of View of the Underdeveloped Countries**

FRANCISCO R. SAGASTI†

1. Introduction

This paper examines the interrelations between underdevelopment, science and technology. As the subtitle suggests, it attempts to present the point of view of the underdeveloped countries with the aim of outlining the basis for a programme of action.

The central thesis is that science and technology were closely related to the emergence of underdevelopment as it has so far arisen in the twentieth century, and that to a certain extent they are contributing to the maintenance and persistence of underdevelopment. In order to modify this situation, major structural changes are required both in the way scientific and technological activities are organized within the underdeveloped countries, and in the international structure of the world scientific and technological effort. These changes will not come automatically. They need to be pressed by the underdeveloped countries themselves. To avoid the disruption and the social costs that may be involved in reshaping their scientific and technological activities, as well as the structure of the world scientific effort, underdeveloped countries should employ the scientific method itself in planning these changes.

The paper begins by examining the concepts of development and underdevelopment, it then explores their relation to science and technology, and ends with some proposals on the contribution that science and technology can make to the development of the Third World.

2. Underdevelopment and Development

Underdevelopment is a phenomenon in its own right. It cannot be adequately studied and interpreted as a "stage" in a sequential development process, or as an

* This paper first appeared in *Science Studies*, **3**, 47–59 (1973).

† The Author is now on the staff of the Department of Scientific Affairs of the Organization of American States and working with the Junta del Acuerdo de Cartagena. This paper does not necessarily reflect the views of the above organizations.

A preliminary draft of this paper was prepared at the Management and Behavioral Science Center of the University of Pennsylvania. The author acknowledges comments and suggestions made by Professor Russell Ackoff on an earlier version. Discussions held with Professor Eric Trist and with Alejandro Moya from the Organization of American States were also helpful, and greatly appreciated.

interval on a development continuum along which all countries can be placed, and through which all must proceed in order to become "developed". As Furtado[1] and Sunkel and Paz[2] have shown, underdevelopment, particularly in Latin America, is a consequence of the historical process of industrialization in Europe and later in North America. Development and underdevelopment are thus two facets of the same process of expansion of western capitalism beginning in the nineteenth century. This process involved the creation and spread of modern technology and the establishment of an international division of labour with a few more advanced countries producing manufactured goods and a large number of backward countries supplying raw materials and primary commodities. Underdevelopment and development evolved simultaneously; they were and are functionally related and they also interact and condition each other. These two phenomena must, therefore, be understood as interdependent parts of a single system. The key factor differentiating these structures is that the developed, by virtue of its endogenous capacity for growth, became dominant, and the underdeveloped, because of its incapacity for growth, became passive, dependent, and dominated.

These statements summarize the "structural" view of underdevelopment proposed by Latin American economists. They describe accurately the situation in Latin America and many other countries of Africa and Asia, although not all underdeveloped countries can be said to fit this pattern exactly.

Following Perroux[3] and Bravo Bresani[4] it is possible to say that the main characteristics of an underdeveloped country are that it is dominated, disarticulated, and incapable of providing an adequate standard of living for the majority of its population. *Domination* implies that the underdeveloped country does not have a capacity for autonomous decision making, and that it has or exercises little control over its own destiny. External factors, beyond the control of the underdeveloped country, are the main determinants of its economic, social and even political decisions.

Domination is formally defined as a bilateral and asymmetric relationship, irreversible in the short and medium terms, characterized because a change in the dominant unit invariably results in a change in the dominated unit, whereas a similar change in the latter has little or no effect in the former (Bravo Bresani and Sagasti[5]). Domination may thus be considered as an extreme form of dependence. The primary form of domination of underdeveloped countries by developed ones is economic, although it has been shown that economic domination is closely related to cultural

[1] Celso Furtado, *Development and Underdevelopment*. Los Angeles: University of California Press, 1962.

[2] Oswaldo Sunkel and Pedro Paz, *El Subdesarrollo Latinamericano y la Teoria del Desarrollo*. Mexico: Siglo Veintiuno Editores, 1970.

[3] Francois Perroux, *L'Economie du XXème Siècle*. Paris: Presses Universitaires de France, 1961.

[4] Jorge Bravo Bresani, *Desarrollo y Subdesarrollo*. Lima: Moncloa Editores, 1966.

[5] Jorge Bravo Bresani and Francisco Sagasti, Elements for a Development Strategy in the Context of the Global Industrial System. Paper presented at the Research Conference on the Multinational Firm and the Nation State. Philadelphia, Pa.: University of Pennsylvania, May 1971.

and technological domination, with one of them leading to or implying the others (Perroux[6], Thebaud[7]).

Developed countries have been continuously shifting their modes of domination over underdeveloped ones in response to changing conditions and to pressure from them. From the control of raw materials extracted from, and manufactured goods supplied to, the underdeveloped countries, developed countries gained control of a significant share of the industrial production facilities through direct investment, and are now gaining control of the technological know-how required in manufacturing. Such control is acquired through direct investment and through such means as licensing agreements, sale of patents, management contracts, and technical assistance agreements. The primary vehicle through which these various forms of domination are exerted is the international or multinational corporation. The changing nature of domination has been explored fully by Furtado[8] and Dos Santos.[9]

Disarticulation means that the underdeveloped country does not constitute a homogeneous unit from the cultural, economic and social points of view. It is a highly stratified society with little or no interaction among the various strata and with almost no mobility between them. As Delgado has shown for the case of Peru, often some of these strata are more closely related to developed countries, to large cosmopolitan urban centres, and to foreign ways of living, than to other strata within the underdeveloped country. As a result, improvements in socioeconomic conditions generally affect only some segments of the population, leaving the other segments or compartments comparatively unchanged.

The third characteristic, the *incapacity to cover the costs of an adequate standard of living for the majority of the population*, is perhaps the most striking of the three. The majority of the world population, which is concentrated in the underdeveloped countries of the Third World, has a very low and often deteriorating standard of living. Traditional aid and trade measures appear to be insufficient to improve the social conditions of the majority of those living in underdeveloped countries. For example, the Interamerican Development Bank Report on Socioeconomic Progress points out deficiencies in practically all aspects of living conditions for most countries in Latin America. Shortages in housing, high infant mortality, low life expectancy, malnutrition, lack of educational opportunities, and high inequalities in income distribution appear to be "normal" for the majority of Latin Americans. Furthermore, these three characteristics of underdevelopment are being reinforced rather than overcome.

Economic development, from the point of view of an underdeveloped country, can be considered a dynamic process of structural change characterized by three factors:

[6] *Op. cit.*, note 3.

[7] Schiller Thebaud, *L'Analyse des Objectifs de la Politique Scientifique des Etats Occidentaux*. Grenoble: Institut Economique et Juridique de l'Energie, Université de Grenoble, May 1969.

[8] Celso Furtado, *Obstacles to Development in Latin America*. New York: Anchor Books, 1970.

[9] Theotonio Dos Santos, "El nuevo carácter de la dependencia". *La Nueva Dependencia*. Lima: Moncloa Campondónico Editores, 1968.

(1) absolute and sustained economic growth, (2) scientific and technological progress, and (3) social propagation of the effects of (1) and (2) to all sectors of the population (Bravo Bresani and Sagasti).[10]

The first factor, economic growth, traditionally has been considered as synonymous with economic development. Although a necessary condition, it is not sufficient by itself, because economic growth, without technological progress and the propagation of its effects, does not overcome the conditions of domination and disarticulation which characterize underdevelopment.

The second factor, technological progress, refers to the autonomous capacity of a country to generate, disseminate and utilize scientific and technological knowledge in its productive and social processes. However, such progress does not imply rejection of technology imported from abroad. Rather it requires the ability to import, absorb and modify foreign know-how, adapting it to local conditions and even re-exporting it after further development. Only through a process of creating its own scientific and technological capacity can an underdeveloped country be in a position to counteract the effects of technological domination. The third and last factor refers to the distribution of the fruits of economic growth and technological progress to all sectors of the population. This is equivalent to eliminating the disarticulation which characterizes underdeveloped countries, distributing the benefits derived from a development process to the various regions of the country and to all segments and sectors of its society, so as to raise living conditions of all to at least minimally acceptable standards.

3. Some Characteristics of the Relation Between Science, Technology and Underdevelopment

In the past two decades, technology has become increasingly important as a factor in the relations between developed countries, as well as in the relations between developed and underdeveloped countries. Technological progress has been credited with contributing more than any other factor to the economic growth of developed countries, and until very recently resources allocated to research and development in advanced countries have been growing at an exponential rate. In addition, the technological content of manufactured goods has become a major determinant of the patterns of trade between developed countries, and also between developed and underdeveloped countries, as the articles in the book edited by Vernon[11] point out.

Most underdeveloped countries, waging a losing battle with the deterioration of the terms of trade, have seen the technological content of their imports increase steadily. At the same time they have realized that it is becoming more and more difficult to manufacture goods for export competing with firms and countries with

[10] *Op. cit.*, note 5.

[11] Raymond Vernon (ed.), *The Technology Factor in International Trade.* New York: Columbia University Press, 1970.

vastly superior scientific and technological capabilities (Prebisch,[12] Schiavo-Campo and Singer[13]).

Many underdeveloped countries have followed the road of import substitution to begin their industrialization (Latin American countries, for example). In most of these cases each new wave of import substitution generated a demand for increasingly more complex and advanced capital equipment. Such equipment was usually provided from abroad, because their incipient scientific and technological infrastructure had no capacity to provide the know-how needed in productive activities. These conditions have led to an increasing dependence on foreign technology, thereby intensifying technological domination. Herrera,[14] Halty,[15] Sagasti[16] and Sanchez Crespo[17] have studied this phenomenon in detail. Thus in some cases the efforts to begin industrialization have led to increased technological dependence. Furthermore, as Dos Santos[18] emphasizes rapid advances in science and technology in developed countries are making technological domination the major form of domination of underdeveloped countries by developed ones in the second half of the twentieth century.

It has been pointed out, particularly by Thebaud,[19] that a country which does not develop a scientific and technological capacity of its own will necessarily be technologically dependent and dominated by more advanced countries. Given this situation there is a great risk that its enterprises and other productive units, forced to acquire know-how from foreign sources (frequently under unfavourable conditions), may become economically dependent on these sources and dominated by them. Beyond a certain threshold the political and cultural independence of the country may be threatened by these forms of dependence and domination.

Most of the technology available in the world today has been created in the developed countries. It has been estimated by the Sussex Group that fully 98% of the world's expenditures on research and development are made by developed countries. Moreover, the 2% spent by the underdeveloped countries is often misused, allocated to research tasks of little consequence for development, and to research centres with lower productivity than the average in developed countries.

[12] Raul Prebisch, *Transformación y Desarrollo*. Washington D.C.: Interamerican Development Bank 1970.

[13] Salvatore Schiavo-Campo and Hans Singer, *Perspectives of Economic Development*. Boston: Houghton Mifflin Co., 1970.

[14] Amilcar Herrera, *Ciencia y Política en América Latina*. Mexico: Siglo Veintiuno Editores, 1971.

[15] Máximo Halty, *Producción Transferencia y Adaptación de Tecnología*. Washington D.C.: Department of Scientific Affairs, OAS, 1970.

[16] Francisco Sagasti, *Towards a Methodology for Planning Science and Technology in Underdeveloped Countries*. Draft Report presented by The Management and Behavioral Science Center, University of Pennsylvania, to the Department of Scientific Affairs of the OAS. Philadelphia, 1972.

[17] Alberto Sanchez Crespo, *Esbozo del Desarrollo Industrial de America Latina y sus Principales Implicaciones sobre el Sistema Cientifico y Tecnología*. Washington D.C.: Department of Scientific Affairs, OAS, 1970.

[18] *Op. cit.*, note 9.

[19] *Op. cit.*, note 7.

The research and development activity in developed countries is concentrated in a few large corporations or government-backed organizations, as shown by OECD studies[20] and by Cooper and Chesnais.[21] These conditions have led to the formation of R & D oligopolies in almost every branch of economic activity, and particularly in those with high technological intensity.

This concentrates the power to exert technological domination in a rather limited number of firms in the advanced countries. These firms exert virtual monopolies over the supply of technology, particularly in their relation to underdeveloped countries.[22] Moreover, the absolute level of the R & D effort in developed countries is several orders of magnitude larger than that of the underdeveloped countries,[23] and the existence of a cumulative scientific and technological tradition in the developed countries makes it difficult for an underdeveloped country, or even a group of them, to reach the levels of achievement attained by the developed ones in almost every area of science and technology.

As a consequence, much of the existing stock of scientific and technological knowledge is suited to the needs and conditions prevailing in developed countries, which have little in common with those of the underdeveloped ones (technologies that require large amounts of capital, large-scale production, and highly skilled labour provide one example). Needless to say, these hardly correspond to the situations prevailing in underdeveloped countries.

Furthermore, as the Japanese White Paper on science and technology[24] points out, current trends in scientific and technological research will intensify this divergence. Among the characteristic trends in the development of new technology, it is possible to identify movements toward more automated plant equipment using limited but highly skilled labour and toward increases in the economic scale of productive units. There is also a trend toward the development of new synthetic materials, which has potentially dangerous implications for underdeveloped countries, because they depend largely on the export of raw materials and primary commodities for their foreign exchange.

Underdeveloped countries thus face a lack of suitable technological alternatives. Often they are faced with the dilemma of choosing between increasing industrial output by using modern, generally capital-intensive techniques, or foregoing opportunities for increasing output but maintaining relatively higher levels of employment using archaic or obsolete techniques. The lack of viable and efficient tech-

[20] Organization for Economic Co-operation and Development (OECD), *The Overall Level and Structure of R & D Efforts in OECD Member Countries*. Paris, 1964.

[21] Charles Cooper and Francois Chesnais, "La Ciencia y la Tecnología en la Integración Europea". *Integración Politica y Económica*, O. Sunkel (ed.), Santiago de Chile: Editorial Universitaria, 1970.

[22] Constantine Vaitsos, Transfer of Resources and Preservation of Monopoly Rents. Paper presented at the Harvard Development Advisory Service Conference. Dubrovnik, 1970.

[23] Geoffrey Oldham, Paper presented at the Round Table on Transfer of Technology, Lima, February 1971. Washington D.C.: Department of Scientific Affairs, OAS, 1971.

[24] Japanese Government, *White Paper on Science and Technology*. Spanish translation in *Comercio Exterior* (Feb. 1971).

nological alternatives, together with the underdeveloped countries' low scientific and technological capacities, poses a difficult challenge, particularly in view of the explosive conditions of population growth, unemployment and underemployment. In some cases the lack of viable alternatives and the ignorance of the buyer of technology in underdeveloped countries has led to wastage of scarce resources, particularly capital. Heavy investments have too often been made in plants with capacities vastly exceeding the size of the market. In some cases governments in underdeveloped countries have encouraged the proliferation of oversize plants for the sake of achieving "competition" (in accordance with imported conventional economic wisdom) with the net result of higher costs, inefficiency and dilapidation of resources.

The capacity for creating technology, or even for absorbing imported technology, is not present in most underdeveloped countries. As a result of the passive character of their economic growth, their demands for technology have usually been satisfied from abroad, through the import of equipment and through technical assistance by foreign technicians. Because foreign technology was readily accessible, little pressure was exerted on the local scientific community to provide viable technological alternatives, particularly in the manufacturing industries. Industrialization policies, fostering technological dependence, have often accentuated the cleavage between the local scientific community and the technological needs of the country. In consequence underdeveloped countries find themselves incapable of creating and satisfying their own technological needs and even incapable of selecting and absorbing the least unsuitable imported technology from the limited range available. Furthermore, as the studies carried out by Vaitsos[25] and the Organization of American States[26] show, foreign technologies are more often than not acquired under very unfavourable conditions, which include high implicit and explicit costs, and restrictions on its use.

Because the productive sectors of the economy exert little pressure on the local scientific and technological communities, scientists, professionals, and technicians orient themselves toward the international scientific community, choosing research topics in fashion, seeking to contribute to the advancement of science as an international undertaking, and disregarding the research needs of their countries almost completely. As long as these needs are ignored by the local scientific communities, they can only retain their identity by orienting themselves outwards. Thus we find local scientific communities of many underdeveloped countries alienated from their own environment and zealously defending freedom of research and the values of universal science, to the detriment of their potential contribution to the development of their own countries. The well-known phenomena of "brain-drain" is an extreme manifestation of this alienation.

The international scientific community, erring by omission rather than commission, has contributed to increase this alienation. Not enough attention has been given by

[25] Constantine Vaitsos, *Strategic Choices in the Commercialization of Technology: the Point of View of Developing Countries*. Mimeograph (July 1970).

[26] Organization of American States (OAS), *La Transferencia de Tecnología hacia los Países del Grupo Andino*. Washington D.C., 1972.

scientists to the scientific and technological problems particular to underdeveloped countries. Sachs[27] postulates that the Europocentric character of modern western science has had a retarding effect on the scientific efforts of underdeveloped countries. Estimates by the Sussex Group[28] indicate that less than 1% of the total research effort in developed countries, to which the international scientific community is primarily related, is directly relevant to the problems of underdevelopment, although the amount may be of the same order of magnitude as the expenditures by underdeveloped countries themselves. Prestige is conferred on researchers who work on exotic and sophisticated advanced topics, choice of which is often dictated by scientific fashion or novelty. Most of these have nothing to do with the more pedestrian scientific and technical problems faced by underdeveloped countries.

The insistence on the international and universal character of the scientific enterprise, the rejection of any type of imposition in the selection of research topics and areas, and the emphasis on the objectivity of science and its quest for truth, have all been proposed as characteristics of what Polanyi[29] has called the "Republic of Science". These, advocated as indispensable conditions for doing science, have also had some effect on the behaviour of the scientific community in underdeveloped countries. Any attempt to reorientate the local scientific effort, gearing it to the needs of the country, is arduously fought by scientists, who see it as infringing on their rights, jeopardizing the integrity of their undertaking, and tampering with the freedom of research.

Little attention is paid by the advocates of unrestricted scientific freedom to the implicit guidance of the scientific enterprise in developed countries through the manipulation of resources available for research. The fact that investments in research soared after the Second World War, primarily in defence, atomic energy and space, is not taken into account. Scientists, professionals, and technicians, preoccupied with the freedom to choose their own individual research topics, are not generally aware of the fact that the broad directions of scientific progress have been primarily influenced by political, economic and social considerations.[30]

It must be emphasized that scientists in underdeveloped countries have acted in a "rational" way in the process of becoming alienated. Given the lack of effective demand for their services in their countries and the structure of the international scientific community, they could not help, if they were willing to remain as scientists, but choose research topics sanctioned by the international scientific community, for which resources were often more freely available. This is but another instance of the divorce between individual and collective rationality in underdeveloped countries.

The preceding paragraphs have dwelt on the aspects of the present organization

[27] Ignacy Sachs, *La Découverte du Tiers Monde*. Paris: Flammarion, 1971.

[28] Sussex Group, *Science, Technology and Underdevelopment: the Case for Reform*. Draft introductory statement for the United Nations World Plan of Action on Science and Technology. Brighton: University of Sussex, 1970.

[29] M. Polanyi, "The Republic of Science", *Criteria for Scientific Development, Public Policy and National Goals*, E. Shils (ed.), Cambridge, Mass.: MIT Press, 1969.

[30] Oscar Varsavsky, *Ciencia, Politica y Cientificismo*. Buenos Aires: CEAL Editores, 1970.

of scientific and technological activities which do not appear to contribute to economic development in underdeveloped countries. This was done primarily because the literature on this subject has emphasized the positive contributions of science and technology to development. This optimistic view must be tempered by the realization that the science and technology of developed nations is not, in the main, the kind that is required in underdeveloped countries; the part that is required is not usually available under favourable conditions, and if it is, there is often a lack of competence to use it.

This is not to deny that science and technology can and will contribute to development. Only latter-day Luddites would refuse to recognize their potential contribution. During the Second World War Britain and the allied countries which considered themselves in a state of emergency and enlisted the help of the majority of their scientists, made an unprecedented effort to utilize science and technology. This showed what is possible to achieve in a relatively short period of time if an all-out concerted effort is made. There is no apparent reason why a similar mobilization to tackle the problems of underdevelopment would not produce equally spectacular results.

To summarize, science and technology have the potential to contribute, perhaps more than any other single factor, to overcoming the conditions of underdevelopment. However, the present structures of scientific and technological activities in both developed and underdeveloped countries are such that this potential is not being fully realized. Rather, they appear to reinforce, at least partially, the conditions of underdevelopment.

4. Conditions for Making Effective the Potential Contribution of Science and Technology to Development

From the preceding analysis it follows that, if science and technology are to contribute to the development of the Third World, there must be a major transformation in the structure of the world scientific and technological effort. The necessary changes, which require a firm commitment both by developed and underdeveloped countries, can be grouped into three categories:

(1) Modifications in the international division of labour and the orientation of scientific and technological activities at the world level.
(2) Generation of local scientific and technological capabilities in underdeveloped countries.
(3) Full incorporation of science and technology within the scope of development planning efforts.

I shall discuss these in turn.

The first group of changes would include measures to ensure that a larger portion of world expenditures in research and development is allocated to projects relevant to underdevelopment. The Sussex Group[31] and the United Nations[32] have suggested targets for the percentage of research and development expenditures by developed countries that should be devoted to the problems of underdeveloped countries (figures around 1–2% have been mentioned) and for the percentage of total aid from developed to underdeveloped countries that should be oriented towards the generation of local scientific and technological capabilities.[33]

These proposals involve primarily bilateral forms of "science and technology aid". On their own they are not likely to have a major effect on the nature and distribution of world expenditures on science and technology. At best they would succeed in improving research and development in specific fields for some underdeveloped countries. At worst they would make the development of local scientific and technological capabilities more dependent on specific foreign sources.

Another proposal along this line seeks to establish multilateral "Research and Development Funds" at the world and/or regional levels. These funds would operate with their own independent and stable financial sources, which could be provided by special taxes or duties on activities that are realized at the world level (for example, Sachs[34] has suggested that programmes to control sea pollution be financed with taxes levied on ships operating in international waters). Voluntary contributions made by underdeveloped countries and matched by developed countries could provide additional funds, but not replace an independent and stable source of money. However, the availability of funds *per se* does not constitute a guarantee that the science and technology financed with them will have an impact on the development of the Third World. If these funds were administered by scientists from underdeveloped countries alienated from their own environment, and/or scientists from developed countries who are not aware of the complex interrelations between science, technology and underdevelopment, the distortions found at present in the world scientific and technological effort would probably be reinforced.

Changes at the international level should also include increased co-operation between underdeveloped countries in matters related to science and technology. Eventually, more intensive collaboration may pave the road for effective scientific and technological integration. However, experience has shown that co-operation agreements are relatively easy to reach when issues of a purely scientific nature are involved, but when co-operation or integration programmes involve research activities that may have direct economic application—which could benefit a particular country or even a particular enterprise—agreement is much more difficult to obtain. In consequence, scientific and technological integration among under-

[31] *Op. cit.*, note 28.

[32] United Nations, *Science and Technology for Development*, Summary of the World Plan of Action on Science and Technology for the UN Second Development Decade. New York, 1971.

[33] At present it appears that Canada, through its International Development Research Center, is effectively moving in the direction of making science and technology a significant component of its aid programme.

[34] Personal communication.

developed countries is likely to be achieved only in conjunction with economic and political integration.

There is also a need for establishing a pressure group that would act on the international scientific community, motivating scientists to become involved in projects of potential or direct benefit to underdeveloped countries. In the face of increased East–West dialogue (and maybe even agreement on nuclear matters) this could be a task for the Pugwash movement. A new generation of scientists-activists who would take the banner of science and technology for the development of the Third World could inject a new sense of purpose into Pugwash. Failing this, it would be necessary to organize a new pressure group, perhaps within the framework of institutions such as UNCTAD's Group of 77, to press these issues before the international scientific community.

The list of possible actions at the international level could be expanded to include the introduction of a "technological dimension" in the evaluation of projects financed by international development banks or agencies, the establishment of mechanisms to award prestige to scientists working on problems related to development (a Nobel prize has once been suggested), and the introduction of a "development merit" criterion for evaluating proposed research projects (following Weinberg's approach to scientific choice).[35]

The second group of changes requires action at the country level. In consequence these changes must be tailored to the particularities of a given national situation. Underdeveloped countries are not a homogeneous lot and specific proposals to improve scientific and technological capabilities must take into account differences in size, resources, levels of modernization, cultural patterns and many other factors of similar nature. Nevertheless, it is possible to identify some actions that are necessary for the development of a local capacity for doing science and technology.

First, it is necessary that long-term objectives be clearly formulated, defining the "style" of science and technology that the country is seeking to develop, and its relation to the overall economic and social development strategy.

Second, the interactions between science and technology and the economic, educational, political and cultural environment of the country must be considered. The environment affects both the demand for knowledge generated by scientific and technological activities and the possibilities for producing it locally. In particular, the characteristics of the economic system and of many government economic policies contain an array of implicit science policies—which are perhaps more important than explicit policies—and which frequently run against the objectives of scientific and technological development. For example, the implicit science policy contained in the import-substituting industrialization of many Latin American countries has often reinforced technological dependence. Identifying these contradictions and exposing these conflicts is of primary importance for scientific and technological development. As long as negative implicit policies remain hidden, it will be impossible to remove the obstacles that prevent scientific and technological development.

[35] Alvin Weinberg, "Criteria for scientific choice", *op. cit.*, note 29.

The development of a local institutional infrastructure for science and technology is a third aspect that must be covered. Institutions are the means through which resources are channelled to scientific and technological activities. A wide and well-organized base of institutions constitutes a necessary condition for attaining an acceptable level of scientific and technological development.

Underdeveloped countries cannot expect to excel in all fields of science and technology. Therefore, a fourth aspect to consider is the need for a strategy of specialization to be followed by the generation of a local capacity for science and technology.[36] This implies a choice of domains in which the underdeveloped country would become an advanced centre of knowledge and to which resources would be allocated with priority. In other fields or domains technology could be imported, although it would be necessary to exert control on these imports to avoid the negative effects that often accompany them. However, this strategy does not imply abandoning the support that fundamental science is required to provide; for example a base of scientific activity and qualified manpower for science and technology.[37] A balance must be achieved between the concentration of resources in some domains and the general support that fundamental science must receive.

The last aspect to consider refers to the availability of resources for science and technology. Human, financial and physical resources must be provided beyond the minimum critical mass, particularly in the domains chosen by the underdeveloped country for specialization. The human resources base for science and technology in underdeveloped countries is rather limited, and probably constitutes the main bottleneck for scientific and technological development. To overcome this limitation it is necessary to implement measures that would produce results in the short-term, such as programmes to repatriate qualified personnel working abroad; in the medium term, such as programmes to expose professionals and scientists to advances in modern science and technology; and also in the long term, such as changes in the educational system. Financial and physical resources must also be increased, although there appears to be a greater need for a more rational use of existing funds and facilities for science and technology.

The third group of changes is oriented towards making science and technology an integral part of development planning. The first task consists in making planners and politicians aware of the role that science and technology play in the development process, emphasizing that technological domination reinforces underdevelopment. It is also necessary to show them that development plans and policies often contain negative implicit policies for science and technology, which make development efforts self-defeating in the long run.

The concern for scientific and technological policy-making and planning is of relatively recent origin. As a consequence, there are no proven and accepted methods

[36] Robert Gilpin, "Technological strategies and national purpose", *Science*, **169,** 441–8, 31 July, 1970.

[37] Simon Rottenberg, "The Warrants for basic research", *op. cit.*, note 29.

and criteria that could be recommended and applied with confidence.[38] Therefore, another task of high priority is the development of, and experimentation with, procedures for scientific and technological policy-making and planning. The scientific method itself should be brought to bear in planning the development of science and technology.

The awareness by planners and politicians of the importance of science and technology for development, and the availability of adequate methods for scientific and technological planning, would combine to legitimize the incorporation of science and technology within the scope of overall development planning efforts.

5. Conclusion

This paper has presented a brief review of the interrelations between science, technology and underdevelopment, and of the issues involved in realizing the potential contribution of science and technology to the development of the Third World. Beginning with a definition of underdevelopment and a conceptualization of the development process, it was possible to show the need for major changes in the structure of the world scientific and technological effort. Proposals for action remained at a relatively high level of generality and were offered as an outline to be explored in further detail in a concrete setting.

The three groups of changes described in Section 4 amount to a radical transformation of the world scientific and technological effort. On the basis of past history, it is highly improbable that these changes will come about automatically. If they are realized, they will be the result of purposeful action by those who stand to gain from them. Therefore, this transformation must be pressed and started by the underdeveloped countries themselves.

However, it is rather unlikely that individual country efforts to modify the international division of labour in science and technology, as well as the norms that regulate the behaviour of the international scientific community, will bear fruit in the short and medium terms. In consequence, underdeveloped countries should begin by organizing their own limited efforts, by augmenting their own capacity for doing science and technology, and by joining forces to initiate the transformation. In the last analysis, this is the only way in which the future of underdeveloped countries will be in their own hands.

[38] Reviews of available planning methods for science and technology can be found in: Organization for Economic Co-operation and Development (OECD), *Analytical Methods in Government Science Policy* (Paris, 1970) and Francisco Sagasti, *A Review and Critique of Approaches and Methods Proposed for Scientific and Technological Planning*. Washington D.C.: Department of Scientific Affairs, OAS, 1970.

*On the Inward- and Outward-looking Components of a Scientific Community**

CARLOS M. VARSAVSKY†

WE ARE to discuss the role that international cooperation can play in Science and Development. To me this means that we should discuss how science and scientists in underdeveloped countries can help the development of their countries, and how scientists and science establishments of the developed countries can contribute towards the same goal.

It often happens, however, that when scientists get together to discuss "Science and Development" they end up discussing the *development of science*, that is, schemes are proposed by which individual or groups of "good" scientists (by "good" it is implicitly or explicitly meant scientists who publish in international journals) can be helped to carry out their research in what is often a hostile or, at best, indifferent environment.

I believe we should make a clear distinction between the development of science, a relatively easy task to perform and which, indeed, concerns only the underdeveloped countries, and the use of science to produce development, a far more complicated and relevant problem, which not only affects the underdeveloped countries but those fully developed as well. The training of scientists and technologists in developing countries should be considered only incidentally, as a means to an end. Our basic goal must be to see how those scientists and technologists can contribute to the development of their countries. The logical train of reasoning should run along more or less the following lines:

1. What does a country wish to achieve through development?
2. What actions are required to reach the above goals and what is the part of science and technology in those actions?
3. How many of what kind of scientists and technologists will the country need?
4. How can they best be trained and in which way can international cooperation contribute to their training?

Before we judge the success of efforts such as the Centers of Excellence, or the potential success of an International Science Foundation, we have to answer the questions posed in points 1 and 2 above.

* Presented at the Pugwash Symposium on Science and Development, Stanford University, September 1970.

† Manager, Area for Research and Development, Fate, Casilla de Correo 7, Victoria B, Buenos Aires, Argentina.

The discussion of point 1 brings us to the need of defining "development". A formal definition can be made by means of a simple mathematical analogy. For a given society we can define "level of civilization" (quantity of life) as a complicated function, say $C(p_i)$ of many variables p_i that represent the different components of a civilization. Then, we can define development, δ, as:

$$\delta = \frac{dC(p_i)}{dt},$$

the first time derivative or rate of change of the level of civilization. As stated above, this is only a formal definition, but it is useful in pointing out that δ can assume both positive and negative values: development is not always for the better.

Because the function C is complicated, no efforts have been made, to my knowledge, to calculate its time derivative in a modern society. It may be possible and useful to start doing it for primitive and isolated societies, where the function may be linear and its rate of change close to zero, but it would be extremely difficult to do it for a modern country, even a most underdeveloped one.

Therefore, when talking about development, one makes a sweeping simplification and eliminates from $C(p_i)$ all parameters that are not strictly related to the economics of a society. One then defines *economic development*, δ_e, as:

$$\delta_e = \frac{de(p_j)}{dt} \; (j_{max} \ll i_{max}, \text{ i.e., all } p_j \text{ are contained in } p_i),$$

the first derivative of the function $e(p_j)$ with respect to time. Both $e(p_j)$ and δ_e can be calculated with reasonable certainty.

In the minds of most people, no difference exists between δ and δ_e; in particular, it is implicitly assumed that if the sign of δ_e is positive, the sign of δ will be positive too. Such trivial confusion has led, and will continue to lead, to unhappiness, both on the individual and on the societal level.

If we could determine the function $C(p_i)$, we would have to analyze here this function. This would lead us to consider the social value of scientific research, that is, to consider a classification of research projects on the basis of social desirability, where by social desirability we mean the extent to which this project maximizes δ for a given society.

However, we are restricted to the use of $e(p_j)$, and to a discussion of the way in which science and technology affect the p_j. Since we wish to maximize δ_e, we focus then our attention on the economic desirability of scientific research, keeping in mind that this is a very subtle concept and not such an obvious thing as it may appear at first sight.

When we take economic desirability into account, two things become immediately obvious. First, in principle, scientists can do research on anything that is technically possible (by "technically possible" I mean conceptually and/or experimentally within our reach). Up to now, the scientific communities have done that; but it is evident that not all research that is technically possible is economically advantageous in the foreseeable future. Second, to be economically desirable, a research project has to satisfy a specific need of the country. As it happens, the needs of different countries are very different and, in particular, the needs of highly developed countries are very different from those of the underdeveloped ones—a point that has to be kept very clearly in mind when talking about international cooperation. These two factors restrict the range of activities that can be carried out by research workers who are engaged in helping their countries' economic development. They also tell us something as to how the scientific community has to interact with other parts of society.

Such interaction, following Sabato[1] and Oldham,[2] can be represented by a tetrahedron (Fig. 1).

It is typical of the underdeveloped countries that the arrow leading from the scientific community to the producers of goods and services is missing, while a very strong arrow connects this community with the highly developed part of the world. To explain this phenomenon, we must look into the structure of the scientific community.

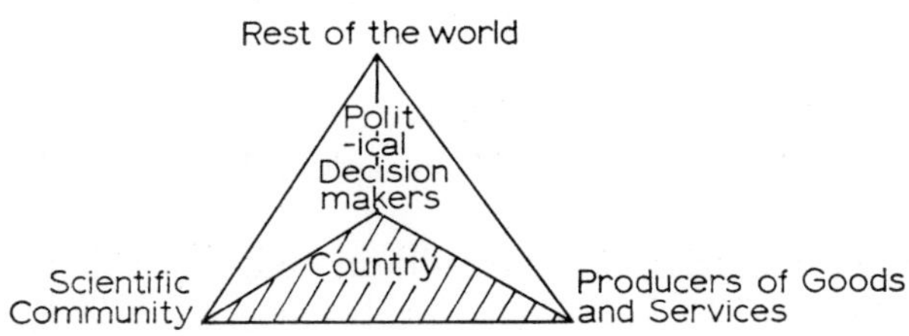

FIG. 1.

A scientific and technological community can be visualized as consisting of two components, of which one will seek its inspiration outside the country and the other within it (both components may coexist in an individual scientist). By this I mean that part of the community will keep up with the latest advances of science and technology in the rest of the world, and will carry out research inspired mostly by such advances, keeping in close touch with those colleagues who work abroad in the same field, and publishing their results in international journals. The other part of the community will look at the national problems and needs for their inspiration and, collaborating from time to time with the outward-looking sector, will try to solve those problems and satisfy those needs.

It is important to point out that inner- and outer-directed research must not be identified with pure and applied research. Both types of research make use of pure and applied science; what distinguishes them from each other is their relevance to the country where the research is carried out.

[1] J. Sabato, *Cuadernos del Centro de Estudios Industriales*, Vol. 1, No. 4, p. 11, Buenos Aires, 1968.
[2] C. H. G. Oldham, private communication.

The overall system is shown in Fig. 2 which includes students:

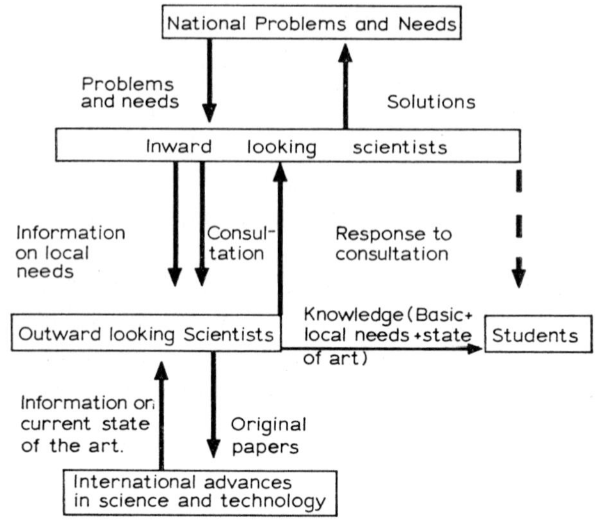

FIG. 2

(the width of the arrows indicates the desirable strength of the correspondent interactions).

In most developing countries, the above-illustrated, desirable system of relations does not exist, the "inward-looking" scientists being practically absent. The lack of linkage between the outward-looking scientists (and, hence, the students) and the national needs alienates these scientists and makes them prone to emigration. The result is the well-known phenomenon of "brain-drain". It is quite obvious that the brain-drain will not be solved by giving the outward-looking scientists better working or living conditions. The only action that will prevent the brain-drain is to establish strong ties between the scientific community and its country. And by strong ties I mean deep involvement of the scientists in the process of economic development.

The Fig. 2 diagram is only qualitative. It still has to be quantified by establishing ratios between the number of scientists that should belong to both parts of the community, and the amounts of money that should be spent on "inner-directed" and "outer-directed" research. We could say that the effort spent on creating each arrow of the diagram should be proportional to its importance. For example, scientists can contribute to the development of their countries even if they sever their ties with the international scientific community (as happened in China); they cannot if they sever their ties with their own countries. Students can be taught by inward-looking scientists; they cannot be taught without danger of alienation by outward-looking scientists alone. Therefore, when resources are limited, priority should be given to the more critical arrows. When discussing the training of scientists in developing countries, and how the international community can assist in their training, we have to recognize that the prime mover, the fundamental box in our diagram is "National Problems and Needs", and that the bulk of the scientific manpower and money should go to the "Inward-looking Scientists".

I stated before that it is characteristic of underdeveloped countries to lack the proper ties between the scientific community and the producers of goods and services. Alternatively, we can say that it is characteristic of underdeveloped countries to lack inward-looking scientists.

When a country has no scientific community at all, outward-looking scientists tend to appear first. Later, to liberate some of those scientists from the grip of world fashion in research, and turn their attention to problems at home, is a difficult step. The international scientific establishment carries much of the blame for this situation.

The proposed International Science Foundation, some Centers of Excellence (such as the Trieste Center) and the work of most foundations seems oriented to ways of strengthening the outward-looking groups of scientists, without proper regard of the fact that such action may be detrimental to the development of the country. To improve the situation, a revision of the policies of the money-granting institutions of the developed countries is required. In addition, greater interaction among the developed countries themselves is urgently needed. In this respect, it may be opportune to quote some paragraphs from the Report of Working Group 4 of the Eighteenth Pugwash Conference (Nice, 1968). The Group recommended the organization of Regional Symposia "in different regions of the developing world", and in reference to them, it said: "These Symposia would also examine the important question: what are the objectives the people of a region have for the development of the region? For example, it may be inappropriate for a particular developing region to adopt in detail or to imitate the methods, values and patterns of life as seen in many of the present advanced countries, and it may be far better for them to work out structures for themselves, consonant with their local heritage and resources. The strategy of development can be defined only after the objectives of development have been defined." ... "Participation in these Symposia should, as far as possible, be confined to scientists of the regions concerned; there could additionally be some recognized specialists in the areas under discussion and some persons long active in the Pugwash Movement to convey the spirit of Pugwash to these Symposia." Later, under the heading "Cooperation between Developing Countries", the Group stated that ". . . too often consideration was given only to relationships between developed nations and developing nations and to aspects of aid, technical assistance, etc., that could be provided by the former to the latter. (We) would like to emphasize, however, that every country in the world can in some respects be in the position of a donor country. It is clear that much closer ties should be established among the developing nations themselves, who are faced with similar situations and problems and are attempting to reach similar goals. International cooperation for peaceful purposes *among the developing nations* can be an important factor in development."

*Science and Development in India—Some Reflections**

YASH PAL†

IN DISCUSSING science and development in India, we are dealing with a country about half the area of the United States, with a population of over 550 million. It is a country where, 20 years ago, only 16% of the population could read and write, and where the present proportion of literacy is only about twice that high. A nearly constant birth rate and a falling death rate, due to improved health measures and eradication of large-scale epidemics, have resulted in population growth rate of over 2.5%. Thus, there are 12 million more Indians every year to feed, clothe, and send to school. The national income of the country was 95 billion rupees in 1950, and has increased to about 330 billion in 1970; however, the prices and the population have also gone up, so that over the last 10 years the real *per capita* income has increased only by about 6%. The absolute *per capita* income at current prices is approximately Rs 600 or about $90 per year. Seventy-five per cent of the population is engaged in agriculture and produce about half the national income.

In juxtaposition with this, one should also look at another aspect of the situation. The country has skills and expertise comparable to those of any country in the world. Its expert manpower includes those who have built and can build large dams and power plants, nuclear reactors and turbines, accelerators and aeroplanes, motor cars and railways, steel mills and oil refineries, micro-wave links and radio-telescopes, radio sets and computers; practically all the wherewithals of a modern technical society. It has economic thinkers of depth, and administrative skills of high order. You find in India great medical schools and large machine-tool factories. There are about seventy national research laboratories. There are over seventy universities and over 2000 colleges, with a total enrolment of about 2 million. There are about 450 engineering schools, turning out about 40,000 engineers every year. The total inventory of natural science and engineering graduates in the country is 250,000 to 300,000 which is nearly half that in the United States. And yet our yearly production of wealth is less than one-twelfth that of the U.S.A.!

Why do all these ingredients not combine to form a living structure—a structure which can take root and become a rapidly growing organism? All the inorganic constituents are present, but the type of organization which characterizes a living organism seems to be lacking. Is the mix wrong? Or do we have to wait for as yet unrecognized catalysts? Or, more likely, is there an already existing organism that

* Presented at the Pugwash Symposium on Science and Development, Stanford University, September 1970.

† Tata Institute of Fundamental Research, Bombay 5, India; currently visiting Professor of Physics at the University of Maryland, College Park, Md.

opposes the growth of the new one? If so, can we tilt the balance against the existing structure and in favor of the new one by a relatively minor change in the mix? For example, it may be argued that even a small deficiency in the number of technical people would have drastic consequences for national economy. That this is not the case in India now is shown by unemployment of some of the best-trained engineers in the country.

As I see it, technical inputs *alone*, either in terms of plant and know-how, or in terms of increasing numbers of scientists and engineers, will not change the development process significantly. The basic problem seems to be social and organizational. As it turns out, what at one time was considered as the major bottleneck—namely the training of scientists and technologists—is a relatively simple task. Substantial steps in this direction already have been taken. There is, actually, a danger that the apparently unproductive nature of this effort may lead to a backlash—that the administrators may be tempted to slow down progress in this direction. What has been found difficult to develop are the social, managerial and organizational patterns which would enable efficient utilization of the available skills and competent manpower. One of the important problems here is lack of self-confidence, particularly on the part of national managers.

We suffer from an underdevelopment syndrome. The traditional belief was that foreign technical skills are always better, that Indian scientists are all right, but that we cannot yet entrust difficult and expensive jobs to them. In general, these attitudes are still with us. As a result, most of our scientists (including engineers) remain unchallenged. This is a vicious circle. Academic school learning remains untempered by the real-life problems. University teachers (train university teachers) who train more university teachers, who train more . . . and this goes on until there are not enough universities to employ them and hence no more jobs. A large part of the technical personnel is insulated from national life. This naturally reflects on the type of students they train.

Many of the trained engineers outside the universities are assigned routine jobs which involve no more than turning knobs and watching dials. Seldom do they encounter the challenge to solve a significant problem. The "underdeveloped" managers do not realize that in applied research the effectiveness and relevance of the scientific effort grows only through continuous interaction with practical problems, that usefulness of knowledge is only discovered through its application. Denied that, much of our research ends up being industrially sterile.

A qualitatively new and bold effort to improve this situation was made by Homi Bhabha, with the strong backing of Jawaharlal Nehru. Bhabha showed that if difficult and challenging tasks are indeed entrusted to young scientists and engineers, they rise to the occasion and grow with the job. Similar vigor and self-confidence are being displayed by a few of the leaders in our scientific enterprises. As Vikram Sarabhai once said, several islands of self-confidence have been created. But these are only islands. Bhabhas and Sarabhais do not run much of the Indian economy. One wonders whether they shouldn't.

Fundamental Science

It is imperative that in order to develop a deeply rooted scientific culture in our country, we must take part in the common drive of mankind towards a better understanding of nature. But does the fact that science is a truly international undertaking necessarily imply that everyone must work at the same point on its broadly advancing front, irrespective of the chances of achievement and consequent satisfaction? Is not the individual or group achievement a necessary catalyst for creativity and further advance?

As it turns out, some of the sectors on the front are more fashionable than others, and most of these fashions are currently set outside India. One can almost say that there is some sort of scientific imperialism in operation, because the very definition of what is good science is determined by what is being done in the U.S., U.S.S.R., or some other countries of the West. For example, a significant fraction of our good theoretical physicists is currently working in the area of particle physics, a branch of science whose great developments during the past decade and a half have been possible mainly because of the very large experimental laboratories set up in Europe, the U.S.A. and the U.S.S.R. We have some excellent centres of particle physics theory in the country. However, it is nearly impossible to have a first-rate development in the particle physics area without experimental thrusts of a magnitude that we can ill afford in India at the present time.

On the other hand, there are several significant areas of scientific endeavour and of great potential advantage to us and to mankind where we do very little. For example very few of our universities even offer any courses in meteorology, or in the broad area of the Earth–Sun relationship.

By its very nature, the quality of fundamental research done anywhere in the world must conform to the highest standards—or it is of no use. This alone suggests the need to select areas of emphasis not only on the basis of fundamental importance, but by taking into account also local advantages in expertise, resources and geographical location. Without significant achievement, the fervour, confidence and atmosphere needed for sustained success cannot be generated. The significant position India occupies in the field of cosmic-ray astrophysics illustrates this situation.

The Green Revolution

One of the most hopeful things that has happened in India (and South Asia generally) is the coming in of the Green Revolution. The fact that food production in 1967–8 jumped upward by 28.8% compared to the previous year, and by about 7.4% compared to the record year 1966–5 has had a tremendous impact. Since agriculture accounts for 50% of India's national product, this represents a substantial increase in India's total wealth production. This has also exploded, rather dramatically, the myth that the Indian farmer is wedded to his old ways and would never accept new agricultural strategies. One of the greatest gains has been that a large fraction of the country's population has come to believe in the relevance of science to solving the practical problem of poverty. However, one must realize that the story of the Green Revolution is not only a scientific success, but also an administrative and managerial triumph, achieved in some selected areas.

I am sure there will be setbacks. New economic disparities and regional imbalances caused by the green revolution will create new problems, but the newly established relationship between science and India's major industry will be a permanent asset.

This is an area where both scientific and managerial strategies have to be largely indigenous. Only loose cooperation with international science is possible, and it must be of a general nature. However, mutual cooperation among countries belonging to the same ecological region could be of immense advantage.

Political Problems and Useless Expenditures

Indian national income is approximately 330 billion rupees, which is about $90 per person. The national budget is about 33 billion rupees. Out of this about 10 billion is spent on defense—about 30% of the national budget and about 3% of the national income. Compared to many other countries, this is not a very large amount; but in relation to the money available for investment and for import of economically strategic materials, it is quite substantial. (It is not known what fraction of the defense budget requires foreign exchange, which is a very critical commodity.) Unfortunately, under the present political situation in this part of the world, it is unlikely that the military budget can be substantially decreased.

There is one aspect of the defense spending in developing countries that makes this expenditure especially pernicious. In the industrial countries, equipping national military forces involves a large industrial effort. The thus generated weapons of war become more and more expensive. When they become obsolete, they can be sold to grateful clients, perhaps even at a profit. Thus industrial production for war feeds, to some extent, the economy—by technological fall-out, if by nothing else. But when *we* buy those weapons, it is money down the drain. This penalty for political animosities is paid year after year.

Whether international science can or should do anything with regard to complex political issues is not clear. But it is certain that these problems contribute greatly to slowing down the pace of development. I ask myself: could it be possible to start a movement called the "India–Pakistan Scientists for Peace and Development"—and have the support of international scientific communities in such efforts? Of course, I am not envisaging slogan-shouting, poster-carrying and liberation of peace doves. The ecologies of our two countries are very similar. Our farmers are dependent on the same vagaries of monsoons. In both countries, a rising water table eliminates enormous amounts of good fertile soil every year. Floods from the same rains, often by swelling the same rivers, sweep away villages in both countries. Cyclones brewing in the Bay of Bengal descend unheralded on the coastal towns of India as well as Pakistan. The sun pours enormous amounts of energy on our parched lands, while we freeze and starve with insufficient fuel resources. We share deserts next to the sea, where groundwater is brackish and where a breakthrough in desalination could make the land capable of producing fruits and flowers. In all these cases, I am not talking about problems of engineering and construction, but of challenging scientific problems. *Just for a start, can we set up a first-rate program aimed at understanding and forecasting the onset and intensity of the monsoons, and in this effort, call for advice and help from our colleagues abroad?*

We read about the havoc wrought by cyclones once or twice every year, but I do not recall a proper advance warning having been ever issued to the endangered areas. In these days of satellite weather watch, could we not jointly set up a cyclone patrol and a warning system?

If we do start on these cooperative scientific ventures, which have a wide human impact, it would be easy to go on to more esoteric and academic areas in which we could do a lot of things together. And maybe this would affect our political relations, too—and decrease the money we must spend on defense, instead of using it for development.

General

This discussion is supposed to deal with what international science can do to help in the development of underdeveloped countries. As I mentioned earlier, in my opinion, our main problem is social and organizational, and there, individual countries have to find their own solutions. If initiatives of the type I just mentioned in relation to India and Pakistan could be taken, cooperation of the international scientific community would become essential.

In a more general way, from the global point of view, the importance of first-rate research in areas such as family planning, solar energy and desalination is much greater than is suggested by the investment of resources and intensity of intellectual effort currently being spent on them.

The next few years are going to see a qualitative increase in the efficiency of communications between different parts of this world. The new means of communication will certainly be used for exposure of until now insulated communities to the commercial and cultural forces of the dominant sections of human society. It is possible that this will also open the way for more effective and productive partnership of the underprivileged sections of human society for their common good. To give a few examples: Is it possible that some of the U.S.-based computer firms will find it profitable to generate the software (and even some hardware) for computers located in populous (and sufficiently educated) areas in South Asia, Africa or Latin America? In a cost study, carried out in connection with planning for bubble-chamber research, it was found that a laboratory in India, analysing about 100,000 events each year, would spend, over a 5-year period, less than one-fifth of what a similar American laboratory does. Properly trained people are available for this kind of work. If communications are good, it is irrelevant where the analyses are done. Are there not many problem areas of this type, both scientific and industrial, where a shifting of certain operations to the developing countries would lead to lowering of the costs of goods and services—in addition to supporting development of these countries? Perhaps this process should be first encouraged at the scientific level—because there, one is relatively free of fears of exploitation and domination.

Lastly, I want to touch on a point which is usually considered to be outside the province of science. When I think of a village in Punjab, which probably has about twice the average *per capita* income of India, I do not see a picture as desperate as would be suggested by the ratio 3300 to 180 between the incomes of an average American and a Punjabi farmer. The Punjabi Jat eats quite well; his children are healthy and go to school; he has adequate clothes; he sings and dances at the village fair; his wife sits at the doorstep on a sunny winter afternoon munching carrot and horseradish and gossips about the dowry neighbors are preparing for their daughter; he owns a bullock cart, maybe a tractor, and certainly a radio, which hangs over his shoulder as he bicycles to the market.

How underdeveloped is this, in terms of the real human condition? By a factor of 20 compared to his American counter? The current definition of human progress in terms of more GNP per person—should it be accepted without question? In view of the problems currently faced by the "overdeveloped" countries, where each solution leads to several new problems concerning the human condition, concerning environment and ecology, should we not question whether it is necessary for the whole of the human society to go through this phase, even if it is much better than that where most people currently are? Is there a possibility of finding new directions for the socio-economic forces?

I do not know what the answers to these questions will be, but we should not become instrumental in perpetuating modern myths about what is needed to achieve a happy and contented society. At least, we should question the prevalent assumptions and think about possible alternatives.

*International Co-operation in the Social and Life Sciences**

THOMAS R. ODHIAMBO†

A NEW enthusiasm is spreading among the political leaders of Africa—the conviction that science and technology can be a major factor in advancing their development goals. This is not something that has suddenly erupted; it is a conviction that has gradually taken root during the last decade as a result of several meetings held in many countries between scientists and political leaders from developing countries. One now hears of a number of government-sponsored experiments being implemented to organize and establish national research councils (for example, in Zambia), of the formation of national academies of science (for instance, in Ghana), and of the appointment of science advisors (particularly encouraged by the UNESCO). These efforts have revealed several problems of major proportions. Firstly, there is an acute dearth of professionally qualified scientists and technologists who can apply existing corpus of knowledge to problems relevant to their countries as well as being competent to create novel frontiers of knowledge. Secondly, the tools for the work in hand—the physical facilities, the equipment, and the workshops for their maintenance, alteration or updating—are woefully inadequate. Thirdly, the sense of isolation is overwhelming for the few scientists that are working in Africa, and the consequent information gap has become a physical reality. Lastly, the environment is not conducive to scientific endeavour: there is the almost audible feeling that "Africa should do things and not think things". The need to solve these problems has consequently brought up the question of international co-operation as a principal technique in meeting the current demands for modernization.

But I want to discuss the question of international co-operation, especially in the life sciences, in a wider context. Science, like love, knows no national boundaries. Indeed, international co-operation in science has become so commonplace over the centuries that it is regarded almost as a truism. I will therefore not spend any time in singing the praises of international scientific co-operation. What I will be much more concerned with is the exploration of new mechanisms for achieving this most desirable objective.

Mechanisms for International Scientific Co-operation

Perhaps the oldest and best tried mechanisms for international co-operation in the scientific field are the following:

* Presented at the Pugwash Symposium on Science and Development, Stanford University, September, 1970.

† Professor of Entomology, University of Nairobi, Kenya; Director, International Centre of Insect Physiology and Ecology, Nairobi, Kenya.

1. The printed book, journal or magazine—which records the research results, the conclusions, and the hypotheses of the authors on some selected scientific problem. In this way, experiences are shared and the information gap is narrowed.
2. The scientific seminar, meeting or conference—which has tended to serve more or less the same objective as the printed book, but has also a different dimension. It has the attraction of providing a platform for scientific rumour-mongering in the best sense of the word, giving participants a chance to learn of the latest and sometimes unsolidified findings from a far-off laboratory. Its greatest value, however, is in facilitating the establishment of close personal contact between scientists of the same general interests.
3. The wandering scholar—and his ability to acquire and transmit information directly. The American physicist, Robert Oppenheimer, was perhaps the most ardent exponent of the wandering scholar in recent times. He believed that the most efficient way to collaborate internationally in science was to let young scientists work in good institutions, let them wrap up the best information inside them, and then encourage them to work in other good institutions elsewhere. Postdoctoral fellowships and allied schemes are intended to do something of this sort.
4. Patenting and the selling of know-how.

These are still potent weapons. Nowhere are they most needed than in the developing countries. And nowhere have I recently seen a more enthusiastic support of these mechanisms for a developing region than in the report by Dr. Glenn T. Seaborg, Chairman of the U.S. Atomic Energy Commission in the course of discussing his 1970 "scientific safari to Africa".[1]

There are other, newer mechanisms that are being tried for their efficacy—not because the older methods no longer work, but for the reason that the pace of scientific advance has become hotter and those that had not started on the historical touch-line with the scientific tradition are likely to be left behind ever more hopelessly.

The newer mechanisms that appear most exciting, in their international implications, are those that have been grouped under the omnibus term of "centres of excellence" or "concentrated research centres" or some similar connotation. These centres are certainly important for the physical sciences as well as the life sciences. But for the latter, they hold a special significance. While the physical sciences are basic sciences in the sense that the order the investigator studies is of a high degree, is more or less predictable, and is subject to mathematical laws, the life sciences have an additional dimension—what we shall loosely call the "life principle"—that disturbs or alters this physico-mathematical order and requires the biologist to be more than a physical scientist. In this situation, the possibility of constant interaction between the physical scientist and the biologist, one variety of a biologist (for example, an ecolo-

[1] Glenn T. Seaborg, "A scientific safari to Africa". *Science*, **169**, 554–61 (1970).

gist) and another variety (for instance, a biochemist), is vitally important. Consequently, it is worth examining these centres of excellence with this possibility in mind.

Centres of Excellence

"Centres of excellence" is a generic term containing under its umbrella at least three main types of concentrated research centres:

1. The model provided by the International Centre for Theoretical Physics, sited at Trieste in Italy—"The Trieste model of Research Centres".
2. The type exemplified by the high-powered, mission-oriented research institutes sponsored by the Ford and Rockefeller foundations; for example, the International Rice Research Institute (at Los Banos in the Philippines), the International Maize and Wheat Improvement Centre (at Chapingo in Mexico), and the newly established International Institute of Tropical Agriculture (at Ibadan in Nigeria). I will call these the "Foundation-type Research Centres".
3. The new model exemplified by the International Centre of Insect Physiology and Ecology (ICIPE) located in Nairobi in Kenya—"The ICIPE model".

I will take these models in turn and discuss their broad objectives and their implications for maximizing international scientific co-operation.

THE TRIESTE MODEL OF RESEARCH CENTRES

One of the most intractable problems of the scientific community in developing countries is the feeling of intellectual isolation and its twin, isolationism. Professionally qualified scientists in any one field are few and far between; they rarely visit each other's laboratories; there are few high-calibre scientific societies where one can test out his intellectual muscles before his peers; funds are extremely hard to come by for visiting advanced laboratories abroad where one can both explain one's research work and derive new sustenance in one's special field; and, as soon as one shows unusual potential as a scientist, he is pushed on to doing unproductive work—administration, committee work, etc. No wonder many scientists in developing countries eventually assume an isolationist trend: they fear, perhaps unconsciously, to expose to the world any inadequacy in their scientific achievement.

The solution—or one of the solutions—to this mammoth problem is to mount a fundamental attack on the impediments on the way to the maintenance of prolonged contact between promising young scientists in developing countries and the best research work going on in the whole world, including that in the developing countries themselves. This is the task that Professor Abdus Salam of Pakistan set for himself to achieve.

In 1962 Professor Salam established the International Centre for Theoretical Physics, at Trieste. At the same time, he created a totally new fellowship scheme of

Associateships.[2] The principal concept behind the scheme is that the Trieste Centre is a permanent meeting-place of young scientists of proved potential and ability, who come from time to time over a period of several years to update their knowledge and find inspiration for their own research programmes back at home. Active scientists from developing countries are granted associateships to enable them to work at Trieste for short or long periods of up to several months, extended over several years, to attend in-depth courses of advanced study in some topic of special interest, to participate in a programme of seminars by distinguished visiting scholars, and to do some quiet thinking on their own current research. The costs for such attendance are met by the Centre; the only obligation on the part of the grantee is that he continues his normal duties as a university teacher or research worker during his periodical returns to his own country.

The strength and significance of this model lies in its providing a sort of travelling fellowship in packages spread over a relatively long period, thus ensuring that the young scientist stabilizes his own programmes at home while feeling that he is very much part of the intellectual world. Indeed, the cross-linkages that have developed among the "graduates" of the Trieste Centre have created the conditions for spreading the intellectual peerage in theoretical physics over a wider sector of the globe than one would have imagined possible even a few years back. The associates have now become an international club of accomplished scientists, whose members are familiar with each other, and who have worked out for themselves an international language of understanding that is likely to prove of potent value in the next phase of the advance of physics.

This model is capable of breaking the back of the problem that Professor Salem set out to solve. It could be repeated in other physicomathematical sciences. I am not certain that it could make such an impact in the biological sciences. The updating of knowledge in the latter, and certainly the communication of worthwhile information, requires that this be done through some form of symbiosis: by imbibing new information, by learning new techniques and procedures through some process of apprenticeship—by actually going and working in a laboratory with your colleagues and scientific peers. The life process is so subject to infinite variations by environmental factors that the search for universal biological laws is not the exciting prospect that it is in the physical sciences.

THE FOUNDATION-TYPE RESEARCH CENTRE

The International Institute of Tropical Agriculture (IITA) was established in July 1968 as an autonomous, non-profit-making, tax-exempt Nigerian corporation. It forms part of an international network of agricultural research institutes sponsored and largely financed by the Rockefeller and Ford foundations. These are the International Rice Research Institute (IRRI) in the Philippines, which over the last 8 years

[2] This is more extensively discussed by A. Salam in *Minerva*, **4**, 461–5 (1966).

has pioneered in the development of new rice varieties and has produced the technology that has led to the "green revolution" now sweeping through Southeast Asia; the International Maize and Wheat Improvement Centre (CIMMYT) in Mexico, which has spread out to many areas of the world new varieties of wheat and maize that have a much higher yield than traditional varieties; the International Centre for Tropical Agriculture (CIAT), now under construction in Cali in Columbia, which will focus its attention on the improvement of tropical pasture and of tropical livestock production; and the IITA, whose major concentration will be the research and socioeconomic studies necessary to ensure the attainment of improved and sustained yields of food crops under tropical humid conditions. This network offers one of the most striking physical realities of the visionary foresight of some of the philanthropic foundations.

The traditional pattern of farming the soils of the humid tropics consists of a system of shifting cultivation, in which the farmer clears an area of its bush and undergrowth and plants it with his crop, from which he gets his living for 2 or 3 years when the soil is likely to become too unproductive for further cultivation. The farmer then opens up a new area, while leaving the old land on fallow for 10 to 15 years, when it is once again ready for a repeat performance. This cultural practice has served tropical peoples well; but with the rising population and widening horizons, the pressure on land has become increasingly great. Consequently, even in mid-Africa where population problems are still minimal, the period of fertility-recovering fallow has steadily contracted, and there has been a consequent drastic decrease in crop yields. To aggravate this situation even further, the high temperature and very humid conditions that persist throughout the year in the humid tropics accelerate the chemical and microbial processes in the soil, and the organic matter is decomposed much more rapidly than in more temperate regions. Leaching of the nutrients is on a large scale; and the torrential rains run off with the best top strata of the soil, leaving lateritic soils that are much more useful for making bricks and plastering walls than for sustaining plant production. Finally, the tropical insects, the luscious weeds, and the plant diseases all combine to make final crop yields only a fraction of what they could otherwise be. In this situation, the use of fertilizers may not only be wasteful—providing as it does more plant material for the insects and the disease to thrive on—but could effectively block the peasant farmer's mind from accepting innovations that do not substantially add to his income.

The IITA has been established with the aim of reversing these trends. Its broad objective is to increase the yield and improve the nutritional content of the basic food crops in the humid tropics. It is also aimed at developing new soil and crop management practices that will stabilize the peculiar soil conditions of the tropics and thus lead to a sustained and permanent cycle of cultivation.

Substantial research has been done on the cash crops one is familiar with in the tropics, such as cacao and oil palm; and these are now a valuable currency in the foreign exchange market for tropical countries in Africa. Relatively little research has been accomplished on the basic food crops of the humid tropics: the protein-rich grain legumes (such as cowpeas and soya beans); the vegetable crops (for instance, egg plant, green beans, and tomatoes); and the root crops (cassava, yams, and sweet potatoes). There are intensive programmes of research on the cereals, such as those at IRRI and CIMMYT and the cereal improvement programmes initiated by a variety of agencies (U.S. Agency for International Development, The Rockefeller Foundation and government ministries) at Serere (in Uganda) and Ahmadu Bello University (in Nigeria). But no such programmes exist for the other major food crops.

The IITA will focus its scientific manpower on the exploration of basic germ plasm, improved genetic combinations, advanced breeding materials, and the most productive and economical breeding methods for these crops. It aims to determine which individual crops are potentially most important in terms of productivity, utility in cropping systems, economic returns, and contributions to improved human nutrition. This is a giant task; and it has made the sponsors of the institute look to its tools for the task in hand.

The sponsors have early recognized the dearth of qualified personnel in the field of agricultural sciences in the tropics. They set its major goal at doing high-quality research work with the purposeful aim of solving a pressing and long-term basic agricultural problem in the humid tropics. But at the same time, the sponsors have taken measures that might well alleviate the shortage of scientific brainpower in Africa by helping to train graduate students as agricultural research scientists, who may then return to the agricultural research organizations in their own countries; by sponsoring regular symposia, workshops, and seminars on various aspects of food agriculture; and by providing a documentation and information centre on soil management and food crop production in the tropics.[3]

The real significance of the IITA model for advanced research is its mission orientation in a relatively narrow field of endeavour. Once having identified the problem area, the sponsors have erected the necessary research plant, endowed it with first-class minds, and lavished on it the equipment and funds necessary to do the job. They have identified the problem, gathered the brains and provided most of the funds. The IITA is sited on a 2,200-acre locality, on which is being erected a water reservoir with a capacity of 350 million gallons for an experimental irrigation programme and research in soil–water relationships. Its staff, when fully developed, will number 400, of which over thirty will be resident senior investigators covering nearly as many specialties and fields, 100 research assistants, and 100 technicians. The capital costs will total something like $17 million; and its annual recurrent costs will amount to about $4 million. In terms of funds annually devoted to research in industrialized countries, these amounts are a mere drop in the ocean. In terms of the developing countries, and in its implications of orienting research talent on problems of relevance to development questions, this type of scientific institute is of paramount importance, far more than the cold figures of total investment convey. It acts as a

[3] Herbert R. Albrecht, "Problems of tropical agriculture: international work on solutions", *Standard Bank Review*, May 1970.

beacon, guiding the search of the newer nations for an efficient scientific tool to use in severing the bonds that tie them to poverty.

THE ICIPE MODEL

In a letter dated 7th October 1966, the Editor of the weekly magazine *Science*, Dr. P. J. Abelson, invited me to contribute a review article on "Science for the Newer Developing Countries" for his magazine. He explained that *Science* had, over the years, published similar articles on other developing areas—in Latin America and Asia—and he thought it might be instructive to have a similar exercise for Africa. The review was to be submitted by 1st February 1967. I took up the challenge as it offered an excellent opportunity to voice the anxieties and frustrations the young scientific community in Mid-Africa was feeling at about that time—the time of the massive attainment of independence, and the time of the equal realization of the massive morass of poverty afflicting these proud new nations.

I will not attempt to review what I said then.[4] I will restate and update the problem as it relates to problems of international co-operation.

A most vital circumstance that one should keep in mind is that some extremely good research was done in colonial times in Africa—in Zaire, in Senegal, in Nigeria, in Ghana, and in the three East African countries (Tanzania, Uganda and Kenya). Indeed, in East Africa, the colonial power left a most impressive infrastructure of agricultural and medical research institutes. However, the foundations on which these were built were extremely weak. I refer to trained indigenous scientists. Hardly any qualified scientists with a productive and creative capacity were left on the laboratory benches at the time of political independence—and few were on the horizon. The university institutions, the cradle of such scholars, was hardly geared to the tempo and condition necessary to nurture the graduate schools that would produce such talented scientists. Consequently, in spite of some outstanding research having been accomplished in the African continent by the colonial power, a little trained scientific manpower was left behind to carry on the good work.

Perhaps even more important is the question of relevance (apart from the question of excellence). How relevant was the research being done to the needs of the country? Had any one surveyed the field and worked out the consequent priorities in research? Who decided the priorities, who were the peers of the scientists working in Africa, and so on?

The biological problems in Africa are vast and likely to inspire one with a sense of frustration—frustration whether in fact the available scientific manpower in Africa will ever make a substantial dent in solving them. In the medical and natural resources fields there are countless large questions awaiting both old and new tools and new hypotheses to tackle them—nutritional problems, parasitological and infectious diseases, psychosomatic disturbances, tropical animal diseases, tropical ecology

[4] Thomas R. Odhiambo, "East Africa: science for development", *Science*, **158**, 876–81 (1967).

and its bearings on questions of the conservation of natural resources, sustained productivity of tropical waters and tropical soils, utilization of forest products, the overwhelming influence of insects on tropical life and history; and one could go on and on listing these major biological stumbling-blocks on the path of African development. Obviously, international co-operation is a necessary factor in the strategy for tackling these problems. Africa left alone with its meager scientific resources (human, physical and financial) will take ages to solve these problems while the rest of the world whizzes past the next phase of development.

It is against this backcloth that the ICIPE was born.

We started with the basic assumption that Africa needs to concentrate its scarce scientific resources in a few very large centres, each with a sharply focused research goal, and each employing a multidisciplinary approach in tackling particular problems over a period of time. In this way, it was possible to see tangible advances being made.

A second basic assumption was that informed Africans themselves must decide the research areas of priority—uninfluenced, in the first instance, by the consideration of foreign centres of advanced research or by foreign scientific experts. This is a principle of cardinal significance. For the ICIPE, I believe, is the first centre of excellence sited in a developing country which has had its basic philosophy for a scientific programme initially decided on by scientists in the developing country itself. The hope therefore is that such an institute is regarded very much as a part of the developing world and is most likely to have a profound influence on development in this part of the world. Having said that, I would like to emphasize that the sponsors of the ICIPE have so designed the ICIPE that in its operations and support it is truly an international centre of advanced research.

The ICIPE was incorporated in April 1970 as a non-profit-making company under the laws of Kenya in such a way that it is "a centre of advanced research meeting the highest standards of world science; that the Centre should have close cooperation with the scientific community in Africa and thus contribute to the development of the latter; that the Centre should have as one of its central aims the making of discoveries and inventions that might lead to the design of new methods for insect pest control; and that the research projects carried out at the ICIPE should capitalize on the wealth of research opportunities that exist in the insect and plant life of eastern Africa."

Insects (and their relatives, such as ticks) are vitally important in the economy and health of tropical Africa (and, indeed, in other tropical regions). The devastations resultant from insect attack on crops and livestock is enormous; the debilitating and hazardous influence of diseases transmitted by insects to man are still serious in Africa. The long-term solution of these problems need a deep knowledge of the biology of the insects concerned. Few research establishments in Africa, by the very nature of their constitution and budgeting, can concentrate their facilities and resources on long-range research. The sponsors of the ICIPE have decided to concentrate all their resources and facilities in long-range research in those aspects of insect ecology and physiology that contain the best promise for pest control without profound side ecological effects.

The central aim of the ICIPE is consequently to carry out fundamental research of high quality with a definite practical goal sharply in focus. Unless such a Centre has certain applied missions which are relevant to the development of Africa, it would have been in danger of becoming yet another academic centre in which individual investigators pursue their own interest, publish erudite articles in scientific journals, but whose efforts do not effectively close the gap between our lack of knowledge of tropical biological problems in this area and the urgent development problems of the region. The sponsors of the ICIPE did not favour the idea of free-wheeling research as the core of the Centre's activities.

A number of innovations in its organization characterize the ICIPE as a unique scientific institute and as one of the finest examples of international co-operation.

1. Its main research workers consist of a pool of young talented postdoctoral researchers in many fields (insect ecology, genetics, organic chemistry, biochemistry, biophysics, physiology, endocrinology, and pathology) from many lands. The criterion for choice is quality of graduate work, potentiality, and talent as a scientist; and this yardstick is used by an international selection panel on candidates from Africa, Europe, North America, Asia, and elsewhere. When the Centre is fully functional it will have, at any one time, thirty or so such workers at the ICIPE Research Centre in Nairobi. These research scientists will be given appointments of 2 to 3 years, and they will then be encouraged to go back to their own home countries to continue their careers. In this way, an international pool of highly informed and motivated scientists of high calibre will be built up in Africa and around the world. Such a scheme of post-doctoral research in developing countries has been graphically described by Dr. Carl Djerassi, Professor of Chemistry at Stanford University, who is one of the pioneering spirits behind the ICIPE.[5]

2. The research workers perform their missions with the guidance of outstanding world scientists who have accepted appointments as Directors of Research. These distinguished scientists have agreed to come to Nairobi two or three times a year to initiate, conduct and advise on research programmes, and to assist the ICIPE Director in working out research priorities and in reviewing the scientific activities of the Centre. At the same time, they have agreed to accept in their own home laboratories talented young African scientists who need further training before joining the ICIPE as research workers.

3. The individual support of the ICIPE comes from over 100 scientists, scientist policy-makers, and other distinguished men of affairs in many countries (Kenya, Uganda, Tanzania, Zambia, Ethiopia, Nigeria, Japan, Australia, Israel, Switzerland, France, Germany, U.S.S.R., Czechoslovakia, the Netherlands, Sweden, U.K., Canada, U.S.A.). They help to give the ICIPE a legal entity.

4. The ICIPE is advised and supported by an International Committee, whose members are nominated by national academies of science and other scientific organiza-

[5] Carl Djerassi, "A high priority? Research centres in developing nations", *Bulletin of the Atomic Scientists*, pp. 22–27 (1968).

tions in many countries. This is a novel device. It is specifically charged, among other things, with guaranteeing the high quality of the work performed at the ICIPE Research Centre in Nairobi. The American Academy of Arts and Sciences and the U.S. National Academy of Sciences, in addition to the East African Academy, the Max Planck Society of the Federal Republic of Germany, the Royal Netherlands Academy of Sciences and Letters, the Israel Academy of Sciences and Humanities, the Royal Swedish Academy of Sciences, the Royal Swedish Academy of Engineering Sciences, the Australian Academy of Science, the Czechoslovak Academy of Sciences, the Japan Science Council, the Centre Nationale de la Recherche Scientifique, and the Office de la Recherche Scientifique et Technique Outre-Mer, were among the first to give the ICIPE tangible and sustained support. Other academies and scientific organizations have shown similar interest and will soon formally join in this international enterprise.

5. The ICIPE is under the supervision of a Governing Board of international composition. Its nine members come from seven countries, with Professor Carroll L. Wilson (Professor at the Alfred P. Sloan School of Management at the Massachusetts Institute of Technology) as its Chairman, and as its Vice Chairman Dr. Wilbert K. Chagula (himself a trained scientist, but until August 1970, Principal of the University of Dar-es-Salaam in Tanzania, and now Minister of Economic Development and Planning in the Tanzanian Cabinet).

Nairobi was chosen as the location of the ICIPE for several reasons. It is an ideal natural laboratory for insect study having an equatorial photoperiod, permitting almost year-round investigations of insects; it has a wide range of climate from tropical monsoon to alpine, and a rich diversity of insects within 200 kilometres of the city. In terms of service and access, Nairobi, with its University and agricultural research organizations, modern city, housing and schools, beautiful countryside and climate, and an excellent worldwide communication system, is a most agreeable place in which to live and work.

The ICIPE is not strictly an applied research institute. Yet, its broad objectives relate to the discovery of knowledge that may fill the critical gaps that now exist in our understanding of the characteristics and biology of the target insect species of great economic importance as pests—tsetse flies, ticks, termites, mosquitoes, to name only a few. Until now, the world has heavily relied for pest control on broad-spectrum chemical poisons which killed all insect species—whether foe (pests) or friend (insects that eat or parasitize other insects)—and did untold damage to the quality of our environment. It did not require deep understanding of insect reproductive physiology, genetics, sensory physiology, population dynamics and the like to determine whether a chemical pesticide was lethal or not. The question was whether the insect keeled over or not when exposed to the chemical. But we now see rapidly accumulating evidence of the severe environmental side effects of persistent chlorinated hydrocarbons. We also observe the increasing resistance of many pest species resulting from mutations in which resistant individuals survived and multiplied.

New avenues for investigations are now called for. Biological control through the release of sterile males is one; but it is only effective in a very few specialized cases. Other avenues are through the use of insect predators, parasites and diseases; but again the potential may be limited and the risks of unintended environmental side effects are large. Yet another avenue is through the selection of species of crops and livestock resistant to insect attack; but this is slow and often the insect pest overcomes the resistance barrier.

The avenue the ICIPE is pioneering in is to seek species-specific controls which do not affect other insects, that do not leave persistent residues, and that have minimal ecological perturbations.

We must fill the gaps in our knowledge in the loop provided by insect ecology–insect physiology–integrated control–environmental quality. We must now understand a great deal about the complex development process through the egg, the larva, the pupa, and the adult insect, a process that is amenable to disturbance or interruption by the introduction of insect hormones at the wrong time, for example. We must have a deeper knowledge of reproductive physiology and genetics, as it offers the prospect of interfering with the mating process or altering the sex ratio among males and females in the offspring. We must possess a clearer understanding of the remarkable sensory organs and the rich system of chemical communication among insect communities of the same species, as it may give us a chance to use false chemical signals to lure insects to their destruction. These and other promising leads of the same nature will be assiduously followed by the band of talented research investigators, trained in many disciplines, at the ICIPE. The general direction of their work will be beaconed by the deep insights and judgements of the senior scientists, who are the Directors of Research at the ICIPE Research Centre.

In addition to the advance of knowledge in these fields, a primary aim of the ICIPE is to bring into a developing region—in this case, Africa—a centre of advanced research where African postdoctoral scientists can find professionally rewarding activity on problems of (a) great importance to the economic development of Africa (and many other parts of the world), but also (b) great intellectual interest. Such a centre of excellence could have a powerful effect on the recognition of science by African leaders as a practical tool, may also encourage African students to study science, and will undoubtedly attract talented African scientists to pursue their profession in their own continent and thus dam the brain-drain before it bursts upon us in earnest.

The University of Nairobi, the Kenya Government, and the East African Community (which links the three nations of Tanzania, Uganda and Kenya in a number of common regional services such as customs, income tax, posts, railways, regional research services and airways) have been quick to see the value of such an institute being sited in Nairobi. The University has provided land for the laboratories of the Centre; the Kenya Government has quickly worked out an agreement which grants the ICIPE a number of privileges (for example, income-tax exemption of the Centre as a whole and the entry of equipment and scientific materials duty free); and the

East African Community is eager to see the ICIPE establish strong links with the applied research institutes concerned with insects, so that as new knowledge is discovered by the ICIPE it may stimulate early field trials and thus decrease the time lag in its application. This last idea is to be expanded to embrace Tropical Africa as a whole. In this respect, an African Advisory Committee is now at work, and will be able to formulate practical measures in the near future.

The vision of the sponsors of the ICIPE is a vast one. Have they the resources to bring it into the realm of reality? The human scientific resources are assured. The Governing Board has assembled together the most distinguished group of insect scientists in the world to act as Directors of Research; the postdoctoral investigators are being appointed on the basis of their outstanding graduate performance and their proved potential as independent research workers; training schemes for graduate technicians are being launched; and the ICIPE Research Centre will rely on trained business managers and other administrative cadres for its day-to-day administration. The Research Centre will be adequately provided with all the equipment that it needs to use for the whole range of its activities. Excellence and relevance are the criteria of the ICIPE in selecting and directing its resources to achieve its goal. Thus, the ICIPE is to possess a first-class research plant, the best brains in the field of its endeavour, and the most efficient administrative machinery for the job in hand.

The question of funding is, however, giving the ICIPE Governing Board a giant-sized headache. Ideally, the ICIPE needs one or two very large financial backers—in the same way that the Foundation-type research centres are being financed. The genesis of the ICIPE, however, and the totally unusual design of the ICIPE has made this problematical. In this situation, the ICIPE Governing Board has turned to international sources of finance and the government-aid agencies as the major sources of finance, with the various academies being responsible for raising funds for some fellowships. The most difficult items on the financial agenda of the ICIPE are capital funds ($1 million), equipment ($2 million), and the operational costs (annual, $1 million) of the Centre. If these can be provided on a long-term basis, then the sponsors will have been given a rare opportunity to demonstrate the utility of this unique experiment in international co-operation and institutional development.

I started this section of my discussion by advancing the possibility that centres of excellence might be of special value as providing the right kind of environment where there is constant interaction between the physical scientist and the biologist. In this way, new streams of thought might be provoked that would otherwise lie unexcited. The ICIPE Research Centre will provide just this kind of cultural environment—where physicists, biophysicists and organic chemists will rub shoulders with ecologists, physiologists and endocrinologists; and where the few carefully selected problems will be investigated from all sides by a whole range of tools. For example, the ICIPE will soon be mounting a long-term research project on insect migration—one biological phenomenon of insect life of great importance in the pest status of certain species. In this case, our studies will be focused on army-worm, a group of moths whose caterpillars devastate a considerable proportion of tropical and subtropical pastureland and agricultural crops every year when these youngsters amass themselves into veritable hordes of feeding machines, mowing down whatever is in their path. A team of ICIPE research workers will be studying the field biology of the adult moths themselves, investigating their flight range, the environmental factors that regulate their seemingly unidirectional mass movement, and the periodical phenomenon of their apparent evidence, using light-trapping techniques, aerial plankton-collecting methods, mobile radar identification procedures, and high-resolution doppler techniques; other workers will be conducting laboratory and wind-tunnel experiments on the biophysics and biochemistry of their flight; while others will be concentrating on elucidating the hormonal and other physiological factors that regulate the migration instinct of the moths; and yet others will be studying the possibility that chemical messengers might be important in the congress of the moths.

Thus, a whole range of biological and physical tools will be brought to bear on understanding the migration biology of the adult moth.

It remains to assess whether the ICIPE model can be applied in other areas of biological research. One cannot pronounce categorically on this at the moment, since it will take at least another 5 years for the full story and potentiality of the ICIPE to unfold. But the iceberg one can see is a marvellous sight; and we are beckoning to the world to join us in the search for what is still to emerge in this crucial historical moment for the oriented and applied development of the biological sciences in the newer nations.

Why Co-operate?

I wish now to turn my attention to a question I begged at the beginning of this discussion. Just how valuable is international scientific co-operation to any particular nation? The question can be answered in many ways by, for example, stating that science does not recognize any national or geographical boundaries, that science is fostered best in an environment of free discussion and consultation, that scientists build on the foundation of discoveries laid by their academic predecessors, that science is an international language that can only develop through all the participating speakers learning of the international rules of the language, and so on and on. A number of truths are imbedded in these assertions, and they have undoubtedly played a role in fashioning science into the pregnant role it is playing in the life of the world today.

But international scientific co-operation has another role to play now. International co-operation in the life sciences can help the new nations to leap-frog into the industrial age quickly, if the new nations judiciously utilize the existing knowledge and techniques in solving their particular problems. The older nations also stand to gain by tackling new questions in new situations using the tools with which they are already familiar. But equally critical to the rest of the world is the probability that such co-operation will bring (from the developing countries) into the international scientific community a new talent, a new genius, a new mode of looking at problems until now shut off from participation and from communing. If this can be facilitated, the world brainpower will then be the richer.

Conclusion

In these pages I have tried to portray the problems and new mechanisms of international scientific co-operation in the life sciences from the point of view of the developing countries, and especially Africa. I have done this for two main reasons. Firstly, because my own knowledge of the world is not that all-encompassing and I wanted to speak where I felt most confidence. Secondly, I am convinced that in formulating scientific policies—whether national or international—one must consider the whole world as his stage. Science is indivisible. A policy decision taken in one country will ultimately affect all other countries, in one way or another. A most gratifying circumstance is that this "natural law" is recognized in the United States; and in their attempt to re-evaluate and re-fashion their national scientific policy they have set out to share their thoughts with other scientists from other countries. This, then, is a beautiful demonstration of the spirit of international co-operation.

Economics and Technology of Development *

A. R. Abdel Meguid†

A Perspective

Over the last two decades the developing countries in Africa, Asia and Latin America achieved on average a rate of growth of output that is arithmetically sufficient to accommodate their growing population and to allow for a modest increase in their living standards. This statistically observed growth did not bring satisfactory progress in development; malnutrition is common, illiteracy and wealth is severely skewed and the gap between rich and poor countries is widening both relatively and absolutely. Even to achieve this modest rise in standards of living, a tremendous mobilization of resources was necessary. In the developing countries of Asia, Africa and Latin America annual investments, during the last decade, amounted to some 15% of these countries' national incomes. On average 80% of these investments was financed from their own resources and 20% from international assistance.[1]

The net flow of financial resources from developed countries (both official and private) to the less developed has been substantial; amounting to some \$120 billion over the period 1956–68. In 1968 the net official and private flow of development assistance has been estimated at over \$12 billion representing 0.5% of the donor countries' gross national product. On average a quarter of this flow was in the form of grants. The financial cost of these capital and technological inputs to the less developed countries, as a result of two decades of borrowing, has been rising sharply. Total recorded foreign debts of the less developed countries stood at \$47.5 billion in June 1968.

The reverse flow of debt service payments on official account amounted to £4.6 billion in 1967. These payments have been rising over the last 10 years by some 17% annually, and are fast approaching a situation where the outflow is equal to the gross flow of new lending. The Pearson Commission estimates that the developing countries of South Asia and the Middle East would reach this position by 1977 and in Africa and Latin America well before that if the flow of lending remains at the 1965–67 level. Even an optimistic increase of 8% annually in the flow of lending

* Presented at the Pugwash Symposium on Science and Development, Stanford University, 1970. The views expressed in this paper are those of the writer and do not necessarily represent the views of the World Bank.

† Secretary of State, Prime Minister's Office, Egypt.

[1] *Partners in Development*, pp. 27–31. Report of the Commission on International Development, 1969.

would mean that between 60% and 90% of new borrowing would be utilized for debt service payments. Export performance of developing countries in the sixties, in spite of annual growth of slightly below 6%, has led to a steady decline from over 30% of total world exports in 1950 to less than 20% in 1968.

The behaviour of these basic variables, which help portray the pattern of international development, leads to justifiable anxiety. Simple and sophisticated extrapolation of recent trends indicate a worsening in the relative position of developing countries, an increasing gap between rich and poor countries at an alarming rate, and a declining support for more aid at less burdensome terms. But, perhaps more important is the observation that despite all the mobilized effort, the economic achievements of the past two decades have by-passed the vast majority of the people in developing countries. This is particularly true of the 60–70% of the population who live in rural areas. But it is also true of the majority of the urban populace. Subsistence wages, unemployment and deteriorating levels of urban services are the results of slow economic growth and rapid migration from the countryside.

It is becoming increasingly accepted that development cannot be a viable process unless and until its social dimension is effectively pursued. A truer definition of development is for a country to achieve a rising but more equitably distributed income, a higher literacy rate, better nutritional standards, a rate of growth of employment opportunities commensurate with the growth of labour force and an improvement in all those economic and social elements which determine the profile of society, its ability to absorb relevant technologies and the quality of its life. These arguments have inspired a dialogue which cautions against the consequences of pursuing these objectives, claiming that they are incompatible with economic growth; that the first priority is to increase the size of national income before worrying about the pattern of its distribution; that the implied development strategy is emotionally inspired rather than realistic, sound and feasible; that the resultant delay in introducing modern technology, though admittedly less labour intensive, would perpetuate the technological gap between advanced and developing countries. In short, the trade-off, it is argued, between the pursuit of economic growth and the achievement of a wide spectrum of social and economic development is large enough to justify a watering-down of social aspirations in favor of purely economic objectives.

One cannot brush aside these arguments for they do contain some isolated elements of truth in varying degrees. They are not, however, based on a penetrating understanding of the totality of the development process. Let us not ask whether growth or social justice provide a more viable approach to development. *Let us ask how to achieve that type of growth whose social impact is both real and widespread.* The international experience of the past two decades provides some important guidelines to the diagnosis and prescriptions for this most relevant question of the 1970s. Few would disagree that the development path has hitherto been uneven, slow and ineffective in terms of both the growth of income *and* the social welfare of the vast majority of the population.

To be candid, in the pursuit of the so-called growth-oriented development strategies, planners have in the past erred in three important respects. *First*, there has been a tendency to assume that the rate of growth is determined almost exclusively by how much a developing country invests. We have viewed development as a process of capital accumulation in the form of new factories, new irrigation works and the expansion of economic infrastructure. The emphasis on new ventures has competed successfully in the bid for governments' attention, for scarce technological, administrative and managerial skills and for the use of domestic resources and foreign financial and technical assistance, to the detriment of existing production capacities. A close appraisal of the economic situation in many developing countries clearly indicates that the near-future growth potentials relate not so much to the need for higher levels of investments as to the need for intensifying efforts to apply newer technologies in order to improve the productivity of the land and the use of the existing capital stock. Enough evidence exists to show that better utilization of established capacities, which require disproportionately less investments per unit of additional output can prove far more productive than the creation of new capacities.

Secondly, there has been an over-reliance by many countries on industrialization to spark the growth process. This approach led to the concentration of industrial investments in urban centres, which were already endowed with some of the required infrastructure, provided ready markets, and had higher political visibility. The industrialization strategy was strengthened by its apparent relative ease and glamour, in contrast to the more obvious difficulties and longer time span involved in modernizing a traditional agriculture, building an extension service and credit and other institutions, educating the farmer, confronting powerful vested landowning interests, and so on.

There have been some success stories of industrial growth, but the failures have been too many. In numerous cases the new industrial urban complex operates inefficiently, at high production cost, low employment impact and excess capacities. Yet profits are artificially jacked up, supported by protection barriers paid for by society to provide entrepreneurial incentives. In most developing countries, the industrial sectors have contributed less than one or two percentage points to the overall growth rate. Nevertheless, industry and its related infrastructural requirements continue to get the lion's share of investment allocations. Economic policies are geared to the industrialization priority, and even government expenditures for social facilities and services are directed disproportionately to urban dwellers. Moreover, the political power of urban agglomerations and organized industrial workers helped increase their share of government attention and of national income. In most of the poorer countries, the increased urban share has been at the expense of the rural sector. But then, the peasantry lack political organization and articulateness, and physically, they are far removed from the centre of political and economic decision-making.

Thirdly, there has been a tendency to overemphasize the adoption of costly modern technology. Intermediate technology, it has been argued, deprives developing countries from the competitive edge they must acquire to lessen their costly dependence on the imports of even their most basic requirements. Of more significance is the emerging costlier financial dependence resulting from the need to import modern know-how and technologies from advanced countries which has overtaxed the system of international technical assistance flows. Foreign experts, many of whom are not sufficiently oriented to the environment of developing countries, are increasingly faced with the problems of helping these countries absorb ill-suited technologies. In many cases, these technologies are designed for differently structured economies and factor-cost relations. The results all too often are much lower yields in terms of output, income and employment than these technologies have produced for the advanced countries themselves.

In short, the pursuit of growth-oriented development strategies has not produced the hoped-for acceleration of economic activities which would enable a subsequent improvement in welfare and equity. Overemphasis on technologically advanced, urban-oriented investments has not proved to be an efficient catalyst for growth, let alone equity and social justice. On the contrary, they have created a tremendous burden on domestic and foreign resources. They have increased the pressures for lower consumption and lower social welfare in order to generate a large enough domestic surplus to provide the required financing. International assistance has flowed generously in support of this effort. But it will have to gather a much faster momentum if it is not to be pre-empted by the reverse flow in honouring the resulting obligations. The scenario, no doubt, can continue for some years. But the impact, in terms of the living standards of two billion people, will remain as marginal as it has been over the past two decades.

A Restatement of the Development Problem

The oversimplified financial approach, dominating the international development dialogue, does not sufficiently take into account the major bottlenecks limiting the full utilization of the natural and the human resource potential of developing countries. The approach is based on a number of economic "identities", of which the following are dominant in the literature and in the philosophy of the existing development institutions:

$$I \cdot K = \Delta Y \quad (1)$$
$$S = I \quad (2)$$
$$S_d + A = S \quad (3)$$
$$Y - C = S_d \quad (4)$$

These identities state that the growth of income (ΔY) is proportional to the magnitude of investment (I) and to the statistically observed output-capital relationship (k); that these investments equal total savings (S) which originate domestically (S_d) as a residual after consumption (C) and are supplemented by foreign aid (A). This approach overemphasizes the magnitudes of the variables without sufficient emphasis on the institutional and other "catalytic" agents necessary to establish the dynamic relations of, and to justify the logic behind, these identities. In other words, the guiding philosophy has implied quantitative rather than qualitative objectives. While the former are certainly significant, the latter have proved to be, in most developing countries, the core of the development problem.

Let us rewrite some of the above identities in "physical" terms.

(1) Given

(a) the scientific assessment of the country's development potential,
(b) the right choice of investment projects,
(c) the successful choice and application of technology,
(d) the economic procurement of machinery,

$\longrightarrow \Delta Y$

(e) the adequacy of infrastructural services,
(f) the availability of production and managerial skills,
(g) the assured supply of raw materials,
(h) the appropriate study of market outlets, . . . etc.,

and

(2) Given

(a) "perfect" financial institutions to mobilize domestic savings and provide investment capital,
(b) unit elasticity of supply of entrepreneurs in the public and private sector,

$S \longrightarrow I$

(c) continuous supply of implementable projects to use up available investment funds,
(d) "acceptable" social and political investment climate,
(e) adequate foreign exchange to finance import requirements, . . . etc.

The neglect of the catalytic agents required to justify the equality sign in our economic identities represent the heart of the problems facing developing countries. There has been a noticeable tendency to implicitly disregard these problems, and concentrate on the magnitude of the variables on both sides of the equations. It is not sufficient for planners to look at the level of their national incomes, to visualize the possibility of a higher level at a future date, to subtract the two magnitudes and to calculate the capital required to bring about the projected increase. Nor is it sufficient for planners to look at their country's balance of payments and to project the trends of exports and imports in order to determine the required level of international support to cover the widening resource gap.

Let us leave the economic planners' offices and seek a better understanding of the causes and the remedies of poverty in villages and towns. This is not an invitation to visit isolated pockets of poverty. But to get closer to understand the way of life of some one and a half billion people, who represent more than one-half of the combined population of the developing world. The majority of this population is in every respect economically and socially deprived. It is their fate which determines the ultimate success or failure of the international development effort.

We need to talk less about the savings–investments gap and more about the life of these people and the urban–rural income gap. We need to embark on the analytical thought process which can deal with the widening technological gap and the rising unemployment gap as effectively as we have developed monetary and fiscal measures to redress the trade gap. We need to agonize less about the trade-off between growth and welfare, and more about the type and content of growth which can truly change the lives of the majority, not just the few. We should not be seeking a strategy of a simple redistribution of incomes by taxing the rich to finance social services and to subsidize the poor. We should be seeking a strategy aimed primarily at the reallocation of new resources in a manner capable of a rapid maximization of production along a growth path that is humanly, not just economically, rewarding.

The present development institutions in the developing countries themselves are not oriented towards the pursuit of quality goals of development with new insights, new strategies and new emphasis. The World Bank has recently spearheaded a concerted effort to expand its role in mobilizing capital and allocating it for enhancing the productive capacity along these lines. The Bank's resources are not unlimited; but the real limitation on the Bank's role, and indeed on all of these "quality efforts", is the identification of the potential areas of action and the appropriate institutional mechanism for their implementation. To overcome these limitations, operational planning requires multidisciplinary institutions to be established within developing countries to work along three main directions: (a) identification and evaluation of exploitable resources, (b) improvement of technology in existing production facilities, and (c) modification of the legal and institutional framework to impart the needed flexibility in implementation.

This task cannot be left to government departments. It should be recognized that in its present form the *institution of government* in most developing countries is a fairly recent phenomenon. Financial and administrative procedures are mostly inherited from colonial administrations and are geared to ensure (often unsuccessfully) against overt fraud in a narrow legalistic-audit sense. Ministries are overburdened by the day-to-day routine of simple governmental operations and the maintenance of law and order. By design and evolution a ministry in most developing countries is a narrowly conceived administrative unit in its early formative stage, not a development and entrepreneurial institution even in the remotest sense. The modern concept of the role of governments, not only as guardians of law and order, but also as active initiators of social reforms and economic development programs has caught many of these countries at the very early stages of transforming a tribal, non-cohesive society into a nation governed under the rule of law administered by a central government—a sufficiently burdensome task.

Moreover, development and productive exploitation of resources must be based on reliable and comprehensive research. This maxim, though widely accepted, is easier to articulate than to apply. Most developing countries have a plethora of resource data, spread among ministries in closely guarded files, much of which is based on rough estimates, informed guesses or partially completed surveys. This welter of often insufficient data seldom provides a realistic basis for sound project preparation and evaluation. The proposed institution at the heart of a country's development machinery should be assigned the task of digesting and assessing the relevance, use and improvement of existing resource data, and of selecting and adapting the most relevant technologies. The entrepreneurial capacity to prepare, evaluate and initiate these investments has been declining, due to social and political factors, in many developing countries. The proposed development institution should participate in the preparation of the complex of projects and decisions necessary to satisfy the wider development objectives. It should be action-oriented and should deal not with intangible aggregates, but with specific technological, social and economic data.

Some Problems of Development in African Cou
Foreign Assistance in Developing Research Cap

F. G. Torto†

This paper will discuss only a few problems; not attempt a comprehe
of the situation in any one country. Further, in view of the internationa
symposium, the discussion will focus on international aspects of the p
more specifically on foreign assistance, particularly by the so-call
countries. The major theme that the author wishes to underline is th
been reached in African development when the main objective of all
tance should be the building-up of African institutions and African con
relevant areas. How to assist in the building of confidence of Africans
of African governments in their own scientists, economists and other
to develop African-based and African-run industries, etc.; these shoul
main objectives of the Africans and of all those who wish to help them.
ment necessary to achieve these objectives is that aid be given to existi
to enable them to develop their full potential, before consideration i
channelling of aid into the establishment of new institutions.

Problems of Scientific Research

Mainly as a result of recommendations made at many conferences h
last decade, and the endorsement of these by the Organization of A
several African countries have recently set up national Science Polic
some countries, the councils are not purely advisory, but have son
control over certain operating agencies, such as research institutes
research related to national development.

Fundamental questions that these countries face are the elaborat
scientific and technological policies, and the identification of approp
by means of which progress will be made towards achieving the desi
It is proper, and even necessary, that the authorities in the develo
should seek to benefit from the experience of more advanced countries
ments in these fields, and various mechanisms could be, and have bee
these stores of experience. These include the foreign adviser or consult

* Presented at the Pugwash Symposium on Science and Development, Star
September 1970.

† Department of Chemistry, University of Ghana, Legon, Ghana.

consultants, whether on purely scientific and technological policies, or on the formulation of an overall development plan. There is much evidence that indicates that countries would stand to benefit by bringing in the foreign element at a stage at which clear guidelines as to the major directions of development have been laid down. Those responsible for the political direction of the country must, in consultation with available local economic, social and technological experts, determine the major technological objectives of the country, based on their own studies of the potential of the country and of the options that are open to them. It is proposed here that unless such clear guidelines have been laid down, the possibility exists that policy will be influenced to an unwarrantable extent by factors such as personal preferences for technological models that are found in advanced countries but which may not necessarily apply to the developing countries concerned, political persuasions of foreign advisers, and even by governmental or commercially dictated vested interests.

Examples of areas in which the wrong decisions could be and have been made, partly under the influence of foreign advisers, are the emphases in higher education, such as the relative claims of elitist academic versus development-oriented technological forms, and the relative merits of prestige research, say in nuclear physics, versus applied research in the utilization of local materials.

Even when reasonable overall policies have been laid down, certain difficulties arise that militate against the realization of the objectives that have been determined. Some of the problems that beset the conduct of research, in research institutes and in universities alike, will be discussed in the following sections.

Personnel for Research

Serious difficulties still arise in many countries from lack of research personnel of all categories. At the highest level, certain of the countries do not have enough persons with sufficient training and experience to provide leadership in the translation of broad policies into detailed research projects, and in the effective direction of such research. This is a possible area for co-operation with more advanced countries; but the solution is not as simple as it might seem. It would not be sufficient, nor is it desirable, merely to send experienced persons from advanced countries to direct research institutes, nor even to help local research directors to formulate programmes. What is really needed in the less developed countries, especially the ones which already have made some headway in the setting up of research laboratories and other scientific institutions and in the training of personnel, is to upgrade their own nationals who show sufficient promise, so as to enable them to occupy senior positions effectively. How to do this is a problem which requires some study.

Possible approaches include intensive training visits for senior personnel from developing countries to developed countries, to gain experience. Another possibility is the organization in a developing country of special workshop(s) of reasonably long duration, say 2 to 4 weeks, which could bring together senior persons from research institutions of different African countries, together with a small number of

experienced and perceptive persons from some developed countries, to study the problems involved in the formulation of relevant research programmes and allied matters. The greatest benefit from such an endeavour is likely to be obtained if it is appreciated that, in view of the far-reaching differences between the conditions obtaining in the typical developing country and those in the average advanced country, the problems under study cannot be solved by the importation of ready-made solutions, patterns and systems from the advanced countries. The aim of the workshop(s) should therefore be an objective and critical examination of the methodology that has proved useful in the developed countries, as a springboard for devising a methodology or methodologies applicable to the specific goals and circumstances of the developing countries.

Small Numbers of Persons of Various Specializations

At a lower level, there are not enough research workers in any one field, whether in research institutes or universities, to permit the formation of teams of sufficient size. This leads to isolation of individual workers and consequent lack of mental stimulation. This is an area in which assistance would be welcomed in many countries. In many instances, the institutions do not have the resources to recruit adequate numbers of persons of various specializations. What is required is, therefore, not help with recruitment, but schemes whereby research workers in selected disciplines could be "loaned" by institutions in the developed countries, for periods of, say 3 years, in order to "catalyse" work in an important field in a research institution in a developing country. It may be pertinent to observe that this is tantamount to sowing "seeds of excellence" in existing institutions, and is, in the opinion of the writer, a better device than creating new "centres of excellence". The latter may become shining examples for a region, but what is an example worth if it cannot be emulated in the many institutions required to tackle the numerous problems that beset these countries.

Lack of Supporting Technical Staff

Another important shortage is that of supporting technical staff. A number of factors contribute to this situation. Thus, the training of technicians in "sub-university" institutions, and the generally inferior conditions of employment that they can enjoy, relegate technicians to positions of inferiority, as regards prestige and economic status in the society. As a consequence, technical training does not attract enough numbers of persons of sufficient calibre to provide the needed cadres of high level technicians. Moreover, facilities for training at the *highest* technical levels are not easy to organize. Among the reasons for this is the large expense of setting up facilities for training a small number of individuals required in certain fields, such as glassblowing.

Two types of action are called for. One is to improve facilities for technical education by the provision of equipment and teachers; the other is to provide various assistance programmes to train promising high-level technicians in the developed countries. A good example of assistance with the establishment of local training facilities is a joint Ghana Government–Canadian Technical Assistance Trade Training Centre in Accra, originally manned by Canadian and Ghanaian staff. After several years the Canadians have withdrawn and left the centre entirely in the hands of Ghanaian staff whom they have trained. The Centre trains technicians in fields such as automechanics, maintenance of refrigeration units, etc., for government and for private firms and organizations. The level of training is not high, but the personnel that the Centre produces is very important for keeping technical services running in many areas important for development.

Maintenance of Equipment—The Spare-Parts Problem

Apart from the relative lack of technicians to service instruments, there is in some countries a serious spare-parts problem. The general level of development of technology is such that there is often only a very small number of equipment of a given type in the whole country. This is particularly true of the more sophisticated scientific instruments, such as spectrophotometers, gas chromatographs, radiation counters and so on. Spare parts for such instruments usually have to be obtained from abroad, and difficulties with foreign exchange and involved ordering procedures can lead to an instrument being out of action for long periods. A useful form of assistance could involve an arrangement whereby small spare parts could be sent quickly from an institution in a developed country, upon request by a laboratory situated, for example, in an African country. This may be an unspectacular form of assistance, but by keeping essential research and other important activities going, it can produce results out of all proportion to the cost involved.

International Experimental Development Centres

An important gap which exists in research and development effort in African countries is the type of organization which can develop an invention, whether it be a process or device or material, from the laboratory scale to prototype or pilot plant stages. Not many African countries can readily assemble the funds and skilled manpower to set up industrial research and development organizations which can cater for all the various needs that are likely to arise. Furthermore, the demand for such development work in any one country, under present circumstances, may not justify the setting up of a large multipurpose centre. The solution that suggests itself is for an international centre serving the needs of several countries. In view of the difficulties which have attended the setting up of international organizations sponsored by a number of African countries, and the discontinuance, at least on an international basis, of some that had been actually started, it would seem that the best hope for the type of multi-purpose centre envisaged here would be to set it up under the auspices of one of the United Nations agencies. Suitable arrangements could be worked out for individual countries to refer appropriate projects to such a centre, and to safeguard the rights of participating countries and of individual inventors.

An arrangement such as the one suggested could help to ensure that African inventions are not stillborn, as they often are at present, and that their exploitation is not left to organizations in developed countries.

Negative Features of Certain Forms of Assistance

Finally, it is necessary to underline the fact that certain forms of, and certain approaches to, assistance and co-operation can have harmful effects on real development of the recipient countries. Real development means progress towards a state of preparedness in the recipient country to carry on satisfactorily after assistance is terminated. To avoid undesirable consequences of assistance, certain conditions must be observed. Some of these are the following:

(i) Visiting experts, and those who serve on advisory and consultative missions, must be persons of really high calibre and experience.

(ii) Visiting experts must spend a long enough period in the country to become familiar with local problems and to give, after a period of "acclimatization", a sufficiently long period of useful service.

(iii) Projects for advisory missions, and eventual implementation of their advice under assistance schemes, should not be conceived entirely by donor countries. They should be seen by local authorities as fitting into their overall development plans. Much harm can be done by foreign countries proposing projects not really vital for the development of the country; such projects often unnecessarily draw upon the recipient country's limited manpower and material resources.

(iv) Projects should not duplicate some which had been already carried out or are under way in the recipient country. There is the example in one African country where surveys for a large irrigation project have been undertaken, within a space of 5 years or so, by no less than four missions from different developed countries. Each mission made recommendations for a project, often repeating those made by previous missions.

(v) Visiting experts, advisory and consultative missions should not hesitate to advise governments to heed the advice they have already received from local scientists and technologists, where such advice is obviously sound. They should refrain from passing the ideas of local individuals as their own. It must be remembered that the building of confidence is, at the present time, a most essential element in the development of African countries and of similar countries or other continents.

*Training of Scientists and Technical Personnel in Developing Countries**

W. K. CHAGULA†

1. Why train local scientists and technical personnel?

The first obvious question which comes to mind, while one is thinking about how to tackle the problem of training local scientists and technical personnel in developing countries, is why we should strive to train local personnel to fill as many of the scientific and technical posts as possible. This is a difficult question to answer, but I am encouraged in my attempt to suggest an answer by the knowledge that the problems which are discussed at Pugwash Conferences are essentially of a political nature and that one of the objectives at these Conferences is to generate new ideas.[1] I am further encouraged to submit my views uninhibited in the knowledge that meetings of Pugwash Conferences are usually private.[1] I shall, therefore, be as frank as possible in the few submissions that I am going to make in this paper.

The two most important requirements in any efficient service, whether technical or non-technical, are continuity and stability. Unfortunately, these are the very ones that are usually endangered immediately after political independence in most developing countries. The best example in recent years, and which will not be excelled for many years to come as an illustration of this fact, is the Congo (Zaire). The facts concerning the Congo (Zaire) are so well known that it would be presumptuous on my part to recount them.

However, let me very briefly state in general terms what happens immediately after the attainment of political independence in any developing country:

(i) A good number of the experienced scientific and technical personnel begin to leave the country usually at very short notice, the process beginning, in some cases, a few months before the actual independence. The reason usually cited for such a mass exodus of expatriate scientific and technical personnel from newly independent countries is that most of them fear the introduction of new laws by the national governments which they believe might adversely affect them, including reprisals against them as a group for any repressive or discriminatory laws enacted by previous colonial governments.

* Presented at the Pugwash Conference on Science and World Affairs, Addis Ababa, Ethiopia, 1965.

† Ministry of Economic Affairs and Development Planning, P.O. Box 9242, Dar es Salaam, Tanzania.

(ii) For a few years after independence, in most developing countries, it becomes very difficult to recruit expatriate technical and scientific personnel, and many essential posts requiring this kind of personnel remain unfilled for long periods of time.

(iii) The few expatriate technical and scientific personnel who do come to developing countries immediately after independence, in most cases do so on very short contracts, leaving the gaps on their departure more or less as they found them on their arrival.

(iv) Many newly independent countries suffer from the side effects of the Cold War, particularly when they actively pursue the policy of non-alignment. Expatriates from certain countries may not agree to come to certain newly independent countries because other expatriates from other countries with a different ideology may happen to be there already.

(v) As a necessary consequence of international politics and intrigue, some key expatriate technical and scientific personnel are, from time to time, declared prohibited immigrants in and expelled from newly independent countries. In some cases, when this happens, there are no local personnel who could replace them immediately.

The above five reasons, and there are many others, I am sure, show clearly that the stability, continuity and efficiency of both the technical and scientific services previously provided by expatriate personnel are almost invariably disrupted in most developing countries immediately after independence. These reasons, coupled with the inescapable fact that the employment of local personnel would be cheaper in the long run,[2] underline the extreme urgency and importance of mounting programmes for the training of local scientific and technical personnel for research institutes, universities, and other technical services in developing countries.

2. For what purpose and to what level should we train local scientists and technical personnel for developing countries?

As a consequence of the arguments I have advanced above under "Why we should train local scientific and technical personnel", it should be obvious that if developing countries are to be as self-sufficient and stable as possible as regards the provision of all their essential services, we should logically aim at training local personnel for *all* the scientific and technical posts that have and will require to be filled in the countries concerned. I shall elaborate below:

(i) *Training for what purpose:*

(a) We shall have to train local university, science, agricultural, veterinary, and engineering teachers to mention only a few.

(b) Local research personnel for the increasing number of research institutes in each developing country will also have to be trained; some of these would be ploughed back into universities as university teachers.

(c) Each developing country is in great and continuing need of science graduate secondary or high school teachers, civil, mechanical, and electrical engineers, doctors, veterinarians, agriculturalists, statisticians, meteorologists and a host of other professional personnel. The social well-being, as well as the creation of the necessary infrastructure to economic development towards which most developing countries are striving, depends primarily on whether there is adequate provision of this cadre of professional personnel, without which political independence would be but an empty and disillusioning objective.

(d) In addition, for each developing country, local scientific and technical personnel to support the top professional personnel and technologists will have to be trained in adequate numbers and to a satisfactory level if the top technical and scientific personnel are to be as effective as they should be. This cadre of supporting staff includes technicians, laboratory assistants, nurses, pharmaceutical assistants, radiographers, agricultural extension personnel, foremen, and other types of supervisory staff. Curiously enough, the lack or insufficiency of this category of supporting personnel does often interfere with the timely implementation of development plans in developing countries, even when there is an adequate provision of the top professional, scientific, or technological staff.

(ii) *Training at what level:*

From the foregoing, it should be clear that the training of local scientific and technical personnel for developing countries will have to be a combined effort of universities, teacher training colleges, technical colleges or institutes and polytechnics, medical schools and nursing training schools, government trade and other training schools attached to Ministries, government and private research institutes and foundations, as well as private industrial and commercial firms—the level of training in each case being determined by the specifications of the posts the trainees will be required to fill later.

3. How and where to train the local personnel

This is a very crucial question on which there is bound to be a divergence of opinion. The important point I should, however, like to make is that the success of any training programme for local technical and scientific personnel would largely depend on the method, or methods, of training selected personnel, as well as on the place or country where the training is done.

In order to stimulate discussion, I shall briefly give my own views on these two crucial aspects of the problem:

(i) Whatever the level of training, whether at universities or other non-university institutions, all local professional, scientific and technical personnel for a particular developing country should ideally be trained in the country concerned so that they could, during their training, be kept in touch with the environment in which they would be working later on. This presupposes that these local training institutions would be adequately staffed and equipped.

The first and most urgent requirement in each developing country, therefore, is to spare no pains either in equipping and staffing the existing training institutions, or in building them from scratch in those countries where no training institutions exist. The help of the relatively developed countries in the task of creating and nurturing these institutions into a stage of maturity and stability with a majority of local personnel running them efficiently would be most valuable.

(ii) In the case of the top personnel for research institutes and universities, although it is true that, for some time to come, the greater bulk of this cadre of personnel will have to be trained in overseas institutions with well-established research and post-graduate facilities, a beginning should be aggressively made *now* in as many developing countries as possible to create and improve the research and postgraduate facilities in their universities and research institutes so that an increasing number of university teachers and research workers could be trained locally.

In this connection, the relatively developed countries could aid the developing countries by offering scholarships and research fellowships to their nationals with the proviso that these scholarships and fellowships should be tenable in universities or research institutes in one of the developing countries.[3] In this way, over a few years, a local "infrastructure" of research expertise and equipment would be built up in the various developing countries.

(iii) Again, ideally, where the necessary training facilities exist, the bulk of the professional personnel for developing countries should be trained locally and active steps should be taken in each country to achieve this objective. This particularly applies to the training of doctors and agricultural professional personnel who should ideally receive their training as close as possible to the environment where they are going to work after graduating. It does also apply to the training of the various categories of supporting technical and scientific personnel mentioned in paragraph 2(i)(d) in this paper.

However, in the case of the personnel for such highly technical professions as engineering and others which could best be studied in highly industrialized countries, while it is still desirable that, as far as possible, they should receive as much of their basic training locally for the reasons I have already given, there should be adequate provision for apprenticeships with appropriate firms or postgraduate training for the graduates before they are enrolled for professional service in their home countries.

The main reason why the apprenticeship or postgraduate training for these categories of professional personnel has to be done in overseas institutions is because hardly any of the developing countries possesses the necessary institutions or well equipped commercial firms for the purpose. With the increasing industrialization of developing countries, however, this problem should become less acute with time.

(iv) Although it is obvious that in the training of local scientific and technical personnel for developing countries the best and most effective methods should be used and that the best and most effective teachers using the best equipment available should be employed, I should stress the importance of each developing country conducting research into all these aspects of training: for training methods that have been found effective in relatively developed countries may produce very disappointing results if applied unmodified in developing countries. Another relevant point I should like to make here is that the most highly qualified teachers on paper may not necessarily be the best and most effective teachers. This point should constantly be kept in mind when selecting expatriate technical teachers for developing countries.

(v) Whether or not the local technical and scientific personnel are trained in local institutions in each developing country, it is important that arrangements should be made for them to attend annual or bi-annual refresher courses or seminars during their employment. Study tours of selected institutions, commercial or engineering firms in the more developed countries should also be of inestimable value to scientists and technical personnel in developing countries, and the assistance and co-operation of the more developed countries in arranging for these study tours would be most valuable.

(vi) Finally, it is usually forgotten that the on-the-job training of *all* categories of scientists and technical personnel should continue throughout the working life of the individuals concerned. This type of training is probably of much greater importance and has much greater potential in developing than in the more developed countries as a source of "high-level" and "middle-level" manpower and should, therefore, be exploited more. I shall touch on this point again when I am discussing the issue "Who should train the local scientists and technical personnel in developing countries?"

4. Who should train the local scientists and technical personnel?

(i) For stability and continuity, as far as possible, and, ideally, a significant core of qualified local teachers should be employed in the various training institutions in developing countries, for local teachers or trainers are much more likely to know the areas of weakness in the backgrounds of the local trainees, and the effective remedial steps that could be taken, than expatriate teachers, however highly qualified the latter might be.

(ii) Local or indigenous trainers or teachers for training institutions usually succeed in surmounting the language difficulty on the part of the students or trainees (most of whom have to use a foreign language) as a barrier to effective communication between the teachers and the students. This language barrier does contribute significantly to the "wastage" of students in training institutions in developing countries.

(iii) The policy-makers and top administrators in the local training institutions should ideally be nationals of the developing country or group of developing countries concerned.

(iv) Once the ideals in sub-paragraphs 4(i) and 4(ii) above are realized, it will always be desirable, and indeed advisable, to have on the staffs of the training institutions in developing countries a small number of expatriate non-indigenous teachers who should be carefully selected for their personal qualities, as well as for their academic or professional standing. However, as far as possible, no non-indigenous or expatriate teachers who have become prejudiced by serving in a position of any importance in a former colonial government should be included on the staffs of scientific and technical training institutions in developing countries. In addition, no expatriate staff who know they stand to lose rather than gain by training indigenous personnel for scientific or technical posts should be recruited for training institutions in these developing countries.

(v) The success of on-the-job training of local scientific and technical personnel, whether at the professional or "supporting" level (see also paragraph 3(vi) in this paper), very much depends on having teams of unselfish, dedicated, unprejudiced, experienced and sympathetic but not paternalistic trainers. All other things being equal, in the long run, qualified and experienced indigenous trainers or teachers should be better than expatriate or non-indigenous ones. As a general principle, those expatriates who will be training themselves out of a job with no other job to fall back upon later should not be employed on the staff for any on-the-job training programme in any developing country. As a corollary to this, the multilateral staff recruiting agreements in which the staff members for on-the-job training schemes have an assured job to go back to afterwards are preferable to the bilateral ones in which the staff members have to start looking for other jobs at the end of their contracts in a developing country.

In conclusion, I should state that some of the views I have advanced in this paper may fall more into the category of "loud thinking" on my part. Nevertheless, these views should not summarily be dismissed as such without serious consideration being given to them.

References

1. Speech by the Secretary-General, Professor J. Rotblat; *Proceedings of the Fourteenth Pugwash Conference on Science and World Affairs*, Venice, 1965, pp. 71–75.
2. G. Burkhardt, "Comments on science education in Africa". *Ibid.*, pp. 112–18.
3. D. G. Osborne and R. H. Wright, *Physics Departments in African Universities*. Quoted by G. Burkhardt in ref. 2.

CHAPTER III

Views from Developed Countries

*Science, Technology and the Developing Countries**

HARRISON BROWN†

THROUGHOUT his million or so years of existence, man has been confronted by countless threats to the survival of his tribes, villages, city-states, nations and civilizations. Hunger, pestilence and violence have continuously taken their toll and throughout history deprivation and misery have probably been the rule rather than the exception. But at no time in history has the prognosis for mankind as a whole been as alarming as it is today.

To observe simply that two-thirds of mankind lives at virtually a subsistence level—ill-fed, ill-clothed and ill-housed—is to take a superficial view of our present condition. The roots of our problems are spread deeply throughout the soil of history and when we examine the trends within our global society as well as its present condition, the prospects indeed appear ominous.

People are tired of reading the warnings of so-called "prophets of doom", and it is not my intention here to be such a prophet. By temperament I am more optimist than pessimist and if I really were to believe that the situation in which man now finds himself were hopeless, I would not bother to write these words. Our situation is desperate, but it is not hopeless.

Before a problem can be solved the nature of the problem must be understood. Before there can be understanding, there must be recognition. One of our basic difficulties today is that the extent of understanding and recognition of our most critical global problems are woefully inadequate to permit meaningful solutions to be found.

The overriding fact which characterizes our world today is that humanity is fissioning into two major groups—the culture of affluence and the culture of poverty. The beginning of the fissioning process dates back to the time, some two centuries ago, when the Industrial Revolution exploded in Western Europe. A succession of technological innovations, beginning with the steam engine, increased fantastically the amount of power that man could mobilize to achieve preconceived goals. The availability of these new levels of power, coupled with a growing body of empirical observations concerning the behavior of materials, gave rise to the greatly increased productivity in industry and agriculture. When systematic research made it possible to couple a growing scientific understanding of nature with the development of new practical technologies, productivity increased still more rapidly.

* Presented to the Committee on Science and Astronautics, U.S. House of Representatives, 92nd Congress. First Session, with Panel on Science and Technology. January 27, 1971.

† Professor of Geochemistry and Professor of Science and Government, California Institute of Technology, Pasadena.

As productivity increased, material wealth increased as well. As wealth increased, health improved, death rates fell and populations increased rapidly—but seldom as rapidly as wealth. The lot of the average individual, therefore, improved substantially in spite of high birth rates. Eventually for reasons which are by no means completely understood, birth rates in these areas fell. When this happened per capita wealth increased even more rapidly than it had previously.

Those nations which were rapidly developing industrially consumed raw materials at increasing rates. Those raw materials that could not be obtained at home were imported. Further, as affluence increased, quantities of products were imported from overseas that were luxuries rather than necessities. Exports were used to pay for the imports and this meant that markets had to be developed. The need for controlling sources of raw materials and markets led to the emergence of the colonial system.

New technologies were transplanted to the colonial areas in a very one-sided manner. For example, on the Indian subcontinent, the development of a network of railroads and the construction of irrigation systems served greatly to decrease the incidence of famine, with the result that population increased substantially. But at no time in most of these areas did increased production exceed increased population to any marked extent. As a result, the lot of the average individual improved with time only very slowly, if at all.

Following World War II the process of fission accelerated, in part as the result of decreasing rates of population growth in the richer countries and increasing rates of population growth in the poorer ones. The process of fission has now reached the point where most people in the world are either very rich or very poor, and not very many people lie in the middle. With each passing year the gap between the rich and the poor nations increases.

Impossible Situation

The culture of affluence is now made up of some 1000 million persons having an average per capita GNP of about $2,300. The culture of poverty is made up of some 2500 million persons having a per capita GNP of about $180, or thirteen-fold less than that characteristic of the rich countries. Within the rich countries per capita GNP is increasing about 4% yearly; within the poor countries the lot of the average person is improving much more slowly—about 2% per year.

As an illustration of the impossible situation that would develop were present growth rates to continue, 130 years would be required for the poorer countries to reach the level of per capita income that is now characteristic of the richer countries. But by that time, at present growth rates, the population of the poorer countries would reach 130 billion persons. At the same time, the per capita GNPs of the richer countries would approach one million dollars per person.

Clearly this is a process which cannot continue indefinitely. Almost certainly the world is slated to experience dramatic, even explosive, changes within the next two or three decades. Modern technology has linked the peoples of the world into a vast communications network. The inhabitants of even the most remote Andean villages listen to transistor radios and an increasing proportion of the world's population has at least occasional access to television. Motion pictures are shown worldwide in even the smallest villages. The poor see how the rich live and would like to share some of the affluence. Desires for more rapid development are intensified, and whereas a few decades ago the poor might have been more or less content with their lot, today they are impatient and are increasingly willing to use strong, even violent, means to help attain their goals. Even in the United States we have seen this phenomenon take place among the poor in our cities. It is a process which is rapidly becoming global, eventually to embrace two-thirds of the human population.

We must add the fact that modern industrial societies, as compared with peasant-village societies, are extremely vulnerable to disruption. We have seen in Indochina how very difficult it is to subdue a determined peasantry even using as we do a broad spectrum of technologically sophisticated weapons. By contrast we have seen how effective the techniques of hijacking, kidnapping and terrorism can be in industrial societies. We have seen large areas blacked out by accidental power failures and many of us, I am sure, have wondered about the potential effect of one or two well-placed bombs in one or two well-selected power stations in the eastern part of the United States in the summer.

The richer nations can, of course, devise increasingly sophisticated techniques aimed at keeping the poor countries from getting too much out of hand. But such techniques are more often than not rendered ineffective by the fact that the rich countries fight among themselves, compete for resources and trade, solicit allies among the poor countries and provide them with massive quantities of armaments. Total military expenditures by the poorer countries in 1969 came to $26 billion. This exceeded by $8 billion the total expenditure by those countries for public education and health.

I suspect that unless development takes place in the future at a rate that is considerably more rapid than at present, the poor countries in their misery will erupt and that the rich countries in their stupidity will take sides.

We see this happening today in several parts of the world and we may well see such developments become widespread in the years ahead. This process could lead to a global war which would drag all of us down, or it could lead to a more or less permanent division of the world into two groups; the rich minority and the miserable majority.

The Development Business

Rapid economic development of the poor countries is essential if we are to create a world in which people world wide can lead lives free of the fears of starvation, disease, deprivation and war. But I am convinced that such a world will not come into existence unless the rich countries take the development business far more seriously in the future than we have in the past. At this stage, it is extremely doubtful that the poor countries can develop by themselves at rates which are truly commensurate with need. Almost without exception they need a great deal of capital and technical assistance from the outside.

The process of development involves an enormously complex array of activities. Resources must be surveyed, analyzed and developed; factories, railroads and highways must be built. In the public sector housing must be developed, educational systems must be planned, teachers and doctors must be trained; in the agricultural sector irrigation systems must be built, varieties of new seeds must be developed, fertilizers and pesticides must be manufactured and delivered to the crops, farmers must be taught new techniques. All of these activities, which are related directly and indirectly to each other, must be carefully planned and executed as parts of a broad development plan. All of these activities require money, and all require cadres of trained people.

Over the years planners have tended to use what is sometimes referred to as the "black-box" approach to development. Here it is assumed that the primary insufficiency in the poorer countries is money and that if only we were to put enough money into one of the black boxes called the "development process", development would automatically flow out of the other end. We now have a great deal of evidence which tells us that this is not correct—that although money is an essential element of development, it is by no means sufficient. Other ingredients must be added, some of which we have come to appreciate only in recent years.

Studies of the economic development of the United States tell us that our own development cannot be explained satisfactorily solely on the basis of the classical economic ingredients of land, labor and capital. Something else has made a contribution which has been fully as important as the three classical ingredients combined, and that "something else" appears to be a mixture of such elements as innovation, research and development, education, training and improved management practices. All of these factors have clearly been involved in the amazingly rapid growth of our agricultural productivity. It has been estimated that the return of the investment in agricultural research, development and education over the past 100 years in the United States has averaged something like 100% annually. Such a large dividend (which we did not realize we were reaping at the time) suggests that we did not invest enough effort in such activities.

In recent years my own experiences and observations in a number of developing countries have led me to conclude that expanded transfers of capital from the rich countries to the poor are essential if development is to be accelerated. It is doubtful, however, that a really major increase in capital flow (a factor of 2, for example) could be effectively absorbed at the present time for the reason that there simply are not enough trained persons in the poorer countries who are able to make the decisions that must be made and to solve the problems that must be solved if development is to take place. Nor is there an adequate organizational structure that would permit decisions to be transformed effectively into action or that would permit development problems to be solved systematically. Nor are there adequate numbers of technically trained persons who can carry out the multiplicity of tasks that are essential in even a quasi-technological society. Indeed, this appears to be a really basic limiting factor to the rate of development.

My first conclusion is that the governments of the richer countries should transfer to the poorer ones as much capital each year as can truly be absorbed effectively.

To start, this would amount to significantly more than present levels of transfer, but not overwhelmingly so, and might within a few years build up on the average to something like 1% of the GNPs of the donor countries. This is hardly a level of transfer designed to bankrupt the richer countries, but handled properly it could have a profound effect upon development rates. Were the rich countries today to transfer 1% of their GNPs to the poor countries, this would amount to 5% of the GNPs of the developing nations. It would correspond to about 10% of what the rich countries are now spending on armaments and to about 60% of the annual direct cost of the Vietnam tragedy to the United States alone. We have now traveled along the military pathway for many years and it has failed to lead us to our goal. It is now time for us to travel along the pathway of compassion.

Technical Assistance

Equally important, however, massive technical-assistance programs should be created, of far greater magnitude than any such program attempted before. These programs should be aimed at producing decision-makers, problem-solvers, managers and other technical persons, at developing the organizational framework within which such persons can operate effectively, and also at devising innovative solutions to specific development problems and bottlenecks.

What should be the basic components of an expanded world program of technical assistance? Here we can obtain some concept of the needs if we trace the implications of programs which are aimed at the primary objective of producing adequate quantities of food.

In the greater part of the developing world agricultural productivity is low both in terms of production per man-hour and production per acre. The effectiveness with which solar energy is transformed into food energy in wheat, for example, is usually limited by the availability of plant nutrients such as nitrogen, phosphorus, potassium and water. Yet adding such nutrients in quantity is usually either ineffective or deleterious because the plant itself has been empirically selected over generations as one best able to survive and grow in the local impoverished environment. The plant is more often than not unable to cope with a substantial increase in the availability of nutrients.

An obvious approach to the solution of this problem is to breed varieties of plants that can thrive in the local environments and that can effectively utilize large quantities of nutrients. This has been a major objective of the Mexican wheat program operated by The Rockefeller Foundation and the government of Mexico for over a third of a century and of the International Rice Research Institute in the

Philippines. Promising varieties of plants that have been developed by those projects have been transplanted to local environments and further modified so they can grow most effectively under the local conditions.

Once this task has been accomplished, sufficient seed must be produced to permit planting by farmers on a significant scale. Fertilizers must be produced on an adequate scale and distributed. Often supplementary water supplies must be provided either from wells or irrigation canals. Pesticides must be produced and distributed.

Thus, to obtain higher crop yields one must construct factories, locate raw materials, provide sources of energy, build highways and trucks, railroads and trains. In other words, a significant elevation of crop yields necessitates a significant level of industrialization.

As crop yields increase, new problems emerge. One must always be on the lookout for new pests which might get out of control. As output per man-hour increases less farm labor is needed and the surplus people migrate to the cities. As nutrition improves, infant and child mortality is lowered and the rate of population growth increases. The urban areas, unable to provide jobs for the migrants, become inundated by vast slum areas and the local government finds itself unable to expand community services as rapidly as people move in. On the farm increased production strains storage facilities and distribution systems.

Thus the apparent solution of one problem, that of low farm productivity, can trigger, as in a chain reaction, a variety of new problems, all of which cry for solution. Further, this one illustration emphasizes how inextricably entwined the problems of a society are—indeed so much so that one must necessarily consider systems rather than isolated problems out of context. We encounter the same kind of situation no matter what priority problem we consider, no matter whether it be public health, population growth, urbanization or industrialization.

Clearly a major component of technical assistance should be science and technology—all related to technological transfer, to the solution of national development problems, and to the building of the organizational structure which will make this possible. The basic aim of technical assistance should be to help a developing nation select, adapt and develop technologies (physical, biological and social) that will help it achieve its social and economic objectives. Development is not simply a matter of transferring existing knowledge and techniques. No matter what the country, no matter what the problem that is to be solved, new problems almost always arise that cannot be solved by remote control. They must be solved locally and the chances are that if they are solved at all, they will be solved by local people who know their culture and environment far better than do most outsiders.

NAS Programs

Research, analysis and problem-solving are major keys to development and should consequently be major elements of technical assistance. Here the United States can play a role which in the long run can far transcend our monetary role. With little question we have the highest technological problem-solving capacity in the world. Properly mobilized and oriented and properly coordinated with programs designed by our friends and colleagues both in the developing and developed countries, I have little doubt that a revolution (using the word in its best sense) can be brought about.

During the past decade the National Academy of Sciences has developed a number of bilateral programs with sister organizations in developing countries aimed at strengthening their local scientific-technological problem-solving competence. In these programs we have brought together natural scientists, social scientists and engineers in roughly equal numbers from our two countries to discuss the problems of development, with particular reference to the role of science and technology. The quality of the discussions and the actions which have resulted have far exceeded our initial expectations. We have helped our colleagues create research councils and institutes; we have jointly modified educational approaches; we have effected nutrition recommendations; we have developed guidelines for industrial research. Indeed, as a result of these experiences I am convinced that the scientific-technological community, worldwide, can play a crucial role in the development process in virtually all developing countries.

Here in the United States we have a real problem. How can we best mobilize our own resources (which are considerable) to better expedite development? Beyond that, what should our total level of support be?

The over-all foreign assistance program of the United States has been the subject of much study in recent years, culminating in the report by the Peterson Panel to the President in the spring of 1970. Perhaps the most important recommendation of that panel was to separate capital assistance from technical asssistance and to channel increasing proportions of our capital assistance through international agencies such as the World Bank and the Inter-American Development Bank. Technical assistance would be placed in an International Development Institute (IDI), which President Nixon in his message to Congress on 15 September 1970 indicated he hoped would be created sometime in 1971.

The channeling of capital assistance through the international agencies makes a great deal of sense, for it decreases the temptation to use foreign assistance as a political weapon.

The creation of an International Development Institute makes a great deal of sense, for properly organized it can provide the structure which can truly mobilize the technical genius of our country for an expanded, reoriented development effort. Such an institute, again properly organized, can enable natural and social scientists and engineers to work on a people-to-people basis with their colleagues overseas on development problems. It would enable development-oriented research to be undertaken domestically as well as overseas. It would make full use of our specialists in industry, agriculture and our universities. Above all, it would provide for the kind of long-term continuity which is so essential if programs are to be successful. Here we must keep in mind that there is no such thing as instant development—that we are dealing with a time scale of a generation or more. A quarter of a century of effort went into the making of the so-called Green Revolution.

In our past technical assistance efforts our programs have been handicapped by short-term outlooks. The yearly budget cycle, the 2-year Congressional cycle, the rapid turnover of AID (Agency for International Development) personnel and the fluctuating international political atmosphere have all combined to make the conduct of meaningful long-term programs at best extremely difficult and, more often than not, impossible. Hopefully the International Development Institute will change that situation.

In what ways might the International Development Institute accelerate the development process?

1. It can develop a sound research base for the solution of the critical problems of development that are now being encountered. I include among these problems associated with rapid population growth, the need for increased food supplies and better nutrition, the problems of rapid urbanization and associated with it massive unemployment, lagging industrialization and the difficulties of transferring technology. I believe that all of these problems are soluble, given the right kind of imaginative effort.

2. The Institute, working with appropriate groups overseas, can help develop local scientific-technological competences by helping to create and strengthen the kinds of local organizations that are necessary and by helping to produce the cadres of trained technical people that are needed. Developing countries must build their local capacities to identify and work on their own research priorities.

3. The United States can develop the incentives and means of orienting an increasing proportion of its total R & D effort to development problems. At present, a trivial fraction of our huge R & D effort is development oriented. We can develop research laboratories for the purpose and encourage young people to choose development careers. We can help universities, industries and other research institutions gear themselves for effective technical cooperation overseas on a stable basis.

4. Technical cooperation can be made into a truly collaborative venture, completely eliminating the "father–child" approach or the master–pupil relationship. We know from sad experience that the United States does not always know best.

5. Technical cooperation can be put on a self-sustaining long-term basis, with continuity of support and programs.

What practical results might come out of such efforts? I can imagine that we might, in collaboration with our colleagues overseas, develop a number of new approaches to the problems of population growth. I can imagine collaborative development of a variety of high labor-input, low capital-input industries. I can imagine our jointly developing "mini-industries", suitable for many locations in developing countries. I can imagine the development of a variety of new approaches to transportation problems in these areas, to the problems of low-cost housing, sanitation and public health. I can imagine a rapid expansion of the Green Revolution to include more countries and more crops. I can imagine new educational techniques, new applications of food technology, and the development of new industries which will make maximum use of local resources.

About a year ago AID reorganized its own programs of technical assistance and created a Technical Assistance Bureau. That Bureau, which has a newly formed unit dealing with science and technology, has already demonstrated on a rather small scale what can be done in spite of numerous handicaps. Its experiences suggest that the International Development Institute, properly organized and adequately funded, can revolutionize the development process.

How large need such a program be? Starting out at a present level of about $500 million, now being devoted to technical assistance, I would visualize that in time it could effectively utilize about some $2 billion annually. Although this is large compared with present programs the situation is desperate and major actions are called for. And I suspect further that we would receive for our money a "bang-for-the-buck" unprecedented in foreign assistance efforts.

A High Priority?
*Research Centers in Developing Nations**

CARL DJERASSI

THE question is raised whether a basic research center of an internationally recognized standard of excellence can be created in a country where no such centers exist and where the requisite scientific manpower is not yet available. An affirmative answer is provided by proposing a model which embodies the following three features: (1) an international cadre of postdoctorate research fellows; (2) overall scientific direction by a group of part-time directors from major universities in different "developed" countries; and (3) selection of research areas with a possible ultimate economic pay-off and a maximum multiplication factor—a typical example from Mexico being cited in support of this thesis. Carl Djerassi is Professor of Chemistry at Stanford University and chairman of the Latin America Science Board of the National Academy of Sciences; the ideas embodied in this article were first presented by him at the Seventeenth Pugwash Conference on Science and World Affairs held in Ronneby, Sweden, September, 1967.

FROM the standpoint of scientific development, a "developing" country becomes a "developed" one when original research emanates from it. The eventual consequence of such research is the creation of technological innovations, which may then be utilized in many other countries that have the manpower to accept such innovations, but not the technical ability to create them.

The ability to perform advanced research, be it in a university or other research center, usually appears at the end of the shopping list in educational and technological priorities of developing countries. The priority items are the rapid widening of the elementary educational base (frequently involved with an eradication of illiteracy), followed by the creation of universities and technological institutes whose main purpose it is to create technicians. It is from this pool of technicians that the teachers, civil servants, public health personnel and applied technologists are generated who are indispensable to the operation of any country. The creation of advanced centers for the training of scientists at the doctoral level with principal emphasis on basic research is usually considered a luxury which comes last, because of the pressure of the higher priority items mentioned above. The few research scientists from such a developing country receive their advanced training abroad and upon returning to their home base either find no opportunity to utilize their training or else encounter conditions which are so inferior to the ones to which they became accustomed during their foreign training that they are not prepared to accept them. The result is the well-

* This paper first appeared in the *Bulletin of the Atomic Scientists*, pp. 22–27 (Jan. 1968).

known "brain-drain"—either by emigration or by remaining in the home country but pursuing quite a different occupation. As far as research potential is concerned, this is equivalent to brain-drain by emigration, because with few exceptions a research scientist in an active field cannot be put on ice for a few years and then be expected to operate creatively after such a hiatus.

The so-called high priority items of developing countries—the improvement in indigenous primary, secondary, and lower university education—have one thing in common; they take many years before noticeable improvements can be achieved. It follows then that even if the financial resources were available, the training of a pool of research scientists able to perform fundamental research would occur sometime in the very distant future. If we now consider the ever-increasing tempo of scientific and technological development in the advanced developed countries, then it becomes obvious that the creation of basic research centers in a developing country through the traditional routes and times sequences is virtually a hopeless proposition. This is particularly true in scientific research where there exists only one standard of excellence, namely an international one. A hypothetical statement such as "this is very good chemical research for Kenya, but rather poor for Sweden" is equivalent to saying: poor chemical research is being performed in Kenya.

Without denying the crucial importance of raising the primary and secondary educational standards of a country, I feel that it is indispensable that a parallel effort be made to establish some centers for basic research of internationally recognized excellence for the following reasons:

1. By demonstrating in a few developing countries that internationally recognized basic research centers can be created without waiting for the logical educational and technological development of the country, an example is created, which (with possible modifications) may prove useful in other developing areas of the world.

2. By selecting a research field (typical examples will be given later) which offers ample opportunity for fundamental research and yet may have eventual practical technological consequences, the possibility of generating new industrial developments is presented.

3. By selecting a research field with a maximum eventual multiplication factor (for example, chemical research, which requires microbiological, entomological, and pharmacological evaluations, and leads eventually to clinical and veterinary applications), a fairly rapid and yet logical broadening in the scope of the research effort can be effected.

4. In countries where scientific research offers no status symbol whatsoever, a beginning will be made in this direction so that eventually scientific research may become a desirable career in that society.

5. An opportunity will be provided in the particular country for advanced training in certain research fields without the necessity initially of sending promising young students abroad. Once they have been exposed to serious research in their own country, subsequent foreign training (even in another field) becomes more meaningful and

a return to the home country much more likely, because they will realize that research of an internationally accepted standard can be performed in their home country.

6. One factor in the complicated brain-drain problem—the unavailability of research facilities—will be partially eliminated in certain fields.

7. To raise the overall scientific level of a country or even of one university is difficult and time-consuming. To create selected centers of excellence in a few fields is easier and their existence, in turn, frequently creates a stimulus in many other areas. Furthermore, the local "image factor", though of limited value, should not be ignored, since it is easier to point to a concrete and operating entity (for example, the Rice Institute in the Philippines) than to a statistically significant reduction in the literacy rate of a country.

The problems involved in establishing such a research center are numerous and several will be peculiar to the particular country selected. Nevertheless, some general proposals can be put forward. Before doing so, I would like to offer a concrete example from my own personal experience which has some bearing on the hypothetical case that I shall present later.

Case History: Mexico

The example which I would like to cite is Mexico in 1949, the year in which I, an organic research chemist, moved to that country for a limited period of time. Several universities existed in the country, notably the huge and venerable National University of Mexico. It had a faculty of chemistry, which graduated several hundred chemists (Bachelor of Science level) each year. By American standards, the training and facilities ranged from mediocre to poor, but apparently they sufficed for the type of positions open to most graduates—analytical chemistry in industry, positions in small pharmaceutical laboratories, the nationalized petroleum industry, etc. No graduate training in the American sense of the word was available and no Ph.D. degree had as yet been offered in chemistry from a Mexican institution. Practically no basic chemical research was being performed in the country and no fundamental chemical discoveries with a technological fall-out had been made in Mexico. On the other hand, a sufficient number of technicians were available so that various foreign technological developments could be adapted locally, the manufacture of certain plastics being a typical example.

A few years earlier, Professor Russell E. Marker of Pennsylvania State College, in connection with his fundamental studies on the constitution of steroidal sapogenins, found that certain Mexican plants contained substantial amounts of a substance, diosgenin, and that the latter could be transformed chemically into the female sex hormone progesterone. At that time, progesterone was difficult to manufacture and only a few very sophisticated pharmaceutical companies in Germany and Switzerland were preparing this expensive material for clinical applications. Marker's discovery led to the organization of a small Mexican company, Syntex, S.A., and largely through the efforts of a young European organic chemist, Dr. G. Rosenkranz (trained in Switzerland), the manufacture of some of the key steroid hormones (testosterone, progesterone, deoxycorticosterone, etc.) was undertaken in Mexico with Mexican technicians using diosgenin (from Mexican plants) as raw material. The methods used for the manufacture of these hormones were based more or less on published procedures and to that extent did not involve fundamental new research, but rather the adaptation of existing technology to local circumstances. In 1949 the therapeutic effects of cortisone were discovered in the U.S. and received worldwide publicity. As a result, academic as well as industrial research teams in Europe and the U.S. started to work feverishly on developing new synthetic methods for this important hormone. In the same year, the hormone sales of the small Mexican company Syntex had risen sufficiently that it could afford a research department of its own and it was for this reason that I joined the company as associate director of chemical research. Since steroid chemistry *per se* was not taught in Mexican universities, most of the Mexican technicians were trained on the premises and a number of fellowships were established, whereby Mexican university graduates could do their research in steroid chemistry in the Syntex Research Laboratories. As a matter of enlightened self-interest, the company subsidized the operation of the Institute of Chemistry of the National University so that in the early 1950s, the major portion of the operating budget of that Institute was based on annual donations from Syntex. As much of the advanced research had to be performed by Ph.D. level chemists, who essentially did not exist in Mexico, most of the research chemists were brought from abroad and by 1959 Ph.D. chemists from over a dozen different countries were working at Syntex. Scientific publication of research results in recognized international journals of the U.S. and Europe was encouraged—in fact so much so that by 1959 more scientific publications in steroid chemistry had emanated from Syntex in Mexico than from any other academic or industrial organization in the world. Some of the research results—for example, the first synthesis of cortisone from plant raw materials or the development of oral contraceptives—also gained general publicity, but the important fact to consider in the context of my particular thesis is that in a matter of 10 years, Mexico—a country in which no basic chemical research had been performed previously—had become one of the world centers in one specialized branch of chemistry. Other companies were also formed and by 1960 well over 50% of the world's supply of steroid hormones originated from that country. Other indirect benefits resulted. While virtually no patents on steroids had previously been filed in Mexico, starting in the early 1950s a veritable flood of applications from all over the world were filed in the Mexican Patent Office and this has continued to this day. The Institute of Chemistry of the National University of Mexico, initially supported largely by Syntex and staffed almost exclusively by "graduates" of Syntex, received larger subsidies from government and philanthropic organizations and developed a Ph.D. program of its own. Many of the first doctoral theses dealt with various aspects of steroid chemistry and gained a well-deserved reputation in that field for the university.

The multiplication factor mentioned by me at the beginning of this paper is further demonstrated by the Mexican example in steroid chemistry. The chemical developments required extensive clinical screening and a great deal of clinical endocrinology and steroid biochemistry was stimulated in the country as a result of the presence of a steroid chemical research center in Mexico. The industrial raw material, diosgenin, was derived from a plant collected wild in the jungles of southern Mexico. As a result of the ever-increasing demand for this raw material, extensive botanical surveys of that portion of the country were undertaken and several companies, including one from Germany, established botanical research stations in an effort to develop a new cultivatable crop. Finally, in recent years the utilization of steroid hormones in animals (better food utilization in cattle, estrus synchronization in sheep, etc.) has prompted a great deal of veterinary research all over the world and some of it is now also being performed in Mexico.

Why did these developments happen in Mexico? It was a matter of proper timing, the existence of a domestic raw material and some entrepreneurship. However, the plant material is also abundant in Central and South America, in South Africa, in India, and in China, and if the proper incentive and entrepreneurs had existed at that time in those countries, the same development could have occurred there instead.

The principal reason why I am citing this Mexican example in such detail is that it illustrates the key component of my proposal for the feasibility of establishing basic research centers in developing countries, namely an answer to the question "who will do this research in the absence of the requisite trained manpower"?

The local supply of Ph.D. chemists in Mexico was negligible and virtually all of them had to be imported. Most of them were in their early twenties and had been trained in the major universities of Europe and the U.S. Why were they prepared to move to Mexico? The salaries were just barely competitive with those paid in comparable American laboratories, but certainly not extraordinary. The physical laboratory conditions were good, but again not superior to anything found in the U.S. or Europe. In order to overcome the obvious isolation inherent in working in locations distant from other active centers of research, opportunities were offered to attend scientific meetings abroad. A few leading university professors from the U.S., England, and Israel served as consultants and visited the Mexican laboratory every 3 months. The excitement of doing research in a foreign country is of particular appeal to younger scientists and this is precisely the group that we wished to attract. Even the scientific publicity value—a scientific publication of good quality from a scientifically unknown country is noted by one's peers much more readily than one emanating from Harvard or Oxford—was a positive factor in the considerations of these young scientists to move to Mexico. Finally, they were given the opportunity of having several local technicians helping them in the laboratory and while this initially required personal training (with its own rewards), it enabled them to work with more "pairs of hands" than would have been possible for them in their home countries in the U.S. or Europe.

Establishing a Center of Excellence

Let us now apply some of these conclusions to the problem at hand—the establishment of an internationally recognized center of excellence in some area of basic scientific research in a country which does not possess the trained manpower for performing such research. Some of the problems and possible solutions for them are the following:

1. *Location*

The research center must be close to an international airport. This requirement is not only for the convenience of the part-time directors (see 3 below), but is particularly important for the rapid delivery of equipment, spare parts, chemicals, etc., on a continuing basis. There is no doubt that virtually all such items would have to be imported by air. While the research center might not be part of a university, at least not during the first few years of its existence (see 4 below), it should be located close to one, because of the great impact that it may have on graduate training in that university. The sharing of library facilities should also be of advantage. Suitable housing for the foreign staff (see 3 below), a reasonable climate and some element of political stability are other factors to be considered. A typical but by no means the only example of a location satisfying these criteria is Nairobi, whose airport is served by over a dozen international airlines.

2. *Selection of Research Field*

This subject and the one of staffing (see 3) are the most crucial. Initially, an international ad hoc committee of scientists from different disciplines should present a list of possible research areas, which should meet most of the following conditions: (a) The field should be an active one in which there is a substantial international reservoir of trained young scientists at the doctorate level in order to satisfy the postdoctorate staffing mentioned under 3. (b) Unusually high capital expenditures should not be required, thus excluding fields such as high-energy physics. (c) Eventual long-term technological developments might emanate from such research (novel drugs or insecticides, novel food supplements, new advances in electronics, etc.). (d) Possible local advantages (availability of plant raw material for steroids in Mexico) might exist for conducting such research in a particular country. With such criteria in mind, it would not be too difficult for such a committee to present a few reasonable conclusions. For instance, although X-ray crystallography is now a very active field of research, its dependence on large computers and the unlikeliness of technological fall-out would presumably eliminate such a field. On the other hand, many aspects of oceanography, starting with organic chemical work on novel natural products from marine sources and ranging all the way to fundamental marine biological studies, could be a suitable field of research in the proper geographical environment.

3. *Staffing*

The basic assumption will be made that for all practical purposes no trained research scientists (as distinguished from technicians) are available in that research field in the country under consideration. The key question then is how to find the necessary scientific staff. I would like to propose an approach which has the virtue of utilizing in a directorial capacity some of the leading experts in a given field, who under ordinary circumstances could not possibly be hired for such a position. At least in my own field of organic chemistry, there exists considerable precedent to indicate that the following approach would be feasible.

Let us assume that the specific research area has been selected and that fifteen scientists at the Ph.D. level are required for the initial program. I would suggest that they be divided into at least three groups, and that the research program of each be directed by a part-time director who will spend no more than 7 days three (or possibly four) times a year at the research center. He will be concerned with the overall research program of his group, but will leave the day-to-day supervisory problems to a younger deputy. The five Ph.D. level scientists working in his group should be postdoctoral fellows, who commit themselves to 2-year (perhaps renewable for a third year) appointments. In areas such as chemistry, such postdoctoral appointments are the rule rather than the exception and the manpower pool is an international one. In my own case, postdoctoral fellows from nearly twenty-five countries have worked in my laboratory during the past dozen years and in most American university chemistry departments, well over 50% of the postdoctoral fellows come from abroad.

One may ask why such research fellows would wish to work in a research center in a "developing" country rather than in one of the major university centers in the U.S. or Europe. Judging from my own experience with a small group in Brazil, many young postdoctoral fellows are tempted to spend a couple of years in an exciting and, for them, different area of the world, provided it is in association with a well-known scientist, in order to be valuable for their eventual career placement. I can cite two very distinguished chemists in my own field who direct such research groups by long distance.

Professor R. B. Woodward, a Nobel laureate from Harvard, has a small post-doctorate research group in Basel, Switzerland, and Professor D. H. R. Barton (Imperial College of Science and Technology), one of Britain's most distinguished chemists, has a similar group in a private research institute in Cambridge, Massachusetts. Each of them visits these groups three or four times a year and some of the research results published by these groups have been nothing short of spectacular. Needless to say, they have had no difficulties in finding suitable postdoctorate fellows from all over the world (I personally know of American, Swiss, English, German, and Indian members of their groups) for these small research groups, even though they were working in privately supported, non-university centers.

I venture to predict that no difficulties in such staffing would be encountered if individuals of such caliber were involved in part-time direction of postdoctorate research groups in a center located, for instance, in Nairobi, and similarly that no difficulty would be encountered in finding such part-time directors. (In connection with the problem of persuading potential postdoctorate fellowship candidates to spend 1 to 2 years in such a location under such circumstances, I posed this question to about fifty chemistry graduate students at Stanford University, using East Africa as a typical location. Well over 50% of them indicated that they would be interested in such a position.)

The advantages of such an approach are many-fold. No single foreign country would have a predominance among the part-time directors or the postdoctorate staff and turnover in the latter group would assure no permanent foreign vested interests. On the other hand, within a few years, one would have created an international cadre of scientists, who are familiar with that particular developing country and who may very well retain some link with it even after having returned to their own developed country to pursue their ultimate professional career. Their original contacts with local students and technicians may well lead to increasing opportunity for the latter group to study eventually in one of the major university centers abroad. The youth and relative lack of formality of young postdoctoral fellows (frequently with no dependants or with small children who pose no schooling problems) is in my opinion an enormous advantage in the interaction between local technicians, students, and apprentice research workers, and this foreign group. Ultimately, as qualified research scientists are trained in that country, they will assume permanent positions in the research center, but the flow of foreign postdoctoral fellows should not be cut off, since it will be the most effective protection against scientific inbreeding.

A separate and more permanent administrative staff would have to be provided, but since its scientific level can be very low, no particular difficulty should be encountered in finding such personnel.

4. *Operational problems*

The ability to attract the required part-time directors, who in turn will be responsible for attracting the postdoctorate fellows, is related to the quality of the research facilities. While it is obvious that most advanced research is more easily performed in New York, London or Zurich than, for instance, in Nairobi, it is indispensable that unnecessary complications be eliminated. Complete duty exemption and very rapid importation by air without delays in customs are minimum conditions, since numerous items for research will have to be brought in continuously. First-class equipment, an appropriate scientific library, good physical facilities, and suitable and reasonably priced housing for the foreign postdoctoral staff are essential. The research center may well have a formal connection or in fact be part of a university or government institution, but in any event there should be intimate informal contact with a university if one exists. In some locations (certain Latin American countries immediately come to mind) official separation from an existing university may be desirable because frequently such institutions are bound to have rules, salary structures, and bureaucratic procedures which could not be tolerated in the research center.

5. *Financial Funding*

The Mexican precedent of private funding by local industry is obviously not relevant, because, if the economic incentive were present, such a research center would already be in operation in the hypothetical country under consideration. A 10-year commitment is probably a minimum and it is likely that the bulk of the funds would have to come from abroad. In the absence of knowledge of the specific research field, it is impossible to calculate the cost of capital expenditures. A very rough budget of $500,000 per annum for operating expenses can be arrived at as follows, and it is obvious that 60 to 80 per cent of it would have to be in hard currency (Table 1).

TABLE 1. HYPOTHETICAL ANNUAL OPERATING BUDGET

15 postdoctorate fellows[1]	$150,000
3 part-time directors[2]	20,000
30 local technicians and assistants[3]	100,000
Administrative manager, staff, secretaries, janitors, etc.	50,000
Research operating expenses[4]	100,000
Rent, amortization of equipment, utilities contingency	80,000
Total	$500,000

[1] $10,000/fellow will include salary, round trip travel expenses, plus one trip at end of first year to international scientific meetings.

[2] Three or four annual trips for each director plus expenses.

[3] This budget item may vary tremendously depending upon the geographical location.

[4] Exclusive of major capital expenditure and major instruments.

It would be preferable if the funds did not come from foreign government sources or at least not from one single foreign country. A consortium of private foundations with possible contributions from some major private industries funneled through some international body might provide a feasible mechanism. My own opinion is that the question of funding should not be considered, until some planning on questions 1 through 4 had first been carried out and a more precise budget determined.

Conclusion

There are many problems associated with the establishment of such a research center which I have not considered. Nevertheless, the main question reduces itself to one of whether it is in fact worthwhile to establish one or more centers of excellence for basic research in a developing country, which according to traditional criteria is not yet ready to perform such research with its own manpower and financial resources. In my opinion it is worthwhile to undertake such an experiment, especially when it is coupled with the type of part-time director and postdoctorate fellowship program outlined above under Staffing. This may be the prototype of an international scientific peace corps of small numbers and very high educational caliber, which may prove viable and useful in other areas as well.

*On the Methodology of Foreign Aid to Less-Developed Countries (LDC) with Special Reference to Science and Technology**

HERBERT MARCOVICH and IGNACY SACHS†

About the Aid Concept

Aid or assistance (sometimes nicknamed "cooperation" or "partnership") belongs to the folklore of contemporary international relations described by Myrdal as "diplomacy by terminology".[1] When a giant American Corporation invests 10 million dollars in Europe, this is business, and Europeans fear, a handsome one for the investor. When the same is done in Asian, African or Latin-American countries (euphemistically calling them "developing" countries), this becomes aid to the recipient country, even though the expectation of return must run pretty high to attract the investor, while the capacity of the recipient country to endure a steady outflow of foreign exchange in form of expatriated profits is far less than that of its European counterpart.

The UNO and OECD compute as aid to a "less-developed country" (LDC) all flow of long-term capital, i.e. grants (putting in the same bag relief given to face some natural calamity and all other kinds of donations), credits (irrespective whether they are granted on [concessionary] business-like or on usury conditions), and direct private investment. All these statistical and terminological tricks add to brighten considerably the picture—still not, by any means, a rosy one—of the flow of resources to LDCs.[2] One must take into account that the real amount of profits expatriated from LDC by foreign businesses is hard to assess, that the substantial investment in industrial countries of capital belonging to citizens of LDC escapes all accounting, and, finally, that losses often incurred by LDC due to adverse terms of trade represent another ticklish and therefore gladly neglected statistical item.

With respect to technical assistance, the mess is still worse. Its volume is assessed by the expenditure involved and the number of experts sent to LDCs. Both measures

* Presented at the Pugwash Symposium on Science and Development, Stanford University, September 1970.

† Pasteur Institute, Paris, France.

[1] Gunnar Myrdal, *The Asian Drama*, New York, 1968.

[2] This note is not concerned with military aid. Though understanding the need for such aid in some very specific occasions, we should like to stress that it does add very little, if anything, to the developmental potential of the recipient country, except for some specific technological spillovers. It distorts the pattern of government expenditure and has far-reaching political implications.

are not indicators of success or failure of the operation. Moreover, statistical evidence is utterly lacking about the real cost of transferred technology (since a substantial part of this cost is included in the price of the imported hardware and the inputs subsequently needed for production, not to speak of the secrecy of many contracts and the impossibility of probing into transactions between subsidiaries of transnational firms).

In other words, "aid" is being used to subsume a whole variety of operations and relationships, prompted by various intentions, and often leading both in the donor country and the recipient country to results quite different from those avowed intentions. The range of intentions on part of the donors goes from moral compulsion to assist the poor, and consciousness of guilt urging to redress the wrongs of the colonial past, through pondered desire of reducing the dangerous tensions between the poor South and the prosperous North, all the way to cold-blooded business calculation and political strategy meant to influence the alignment of the LDC's in the worldwide competition of big powers.

All these motivations may be present at once; or a political operation may be presented to the public opinion in the donor country as arising from recognition of ethical and moral duty. As for the governments of LDCs, some accept aid in the present forms and press for more of it, because they genuinely believe that in spite of all shortcomings, it is still better than no aid at all. Others are so entrapped in a pattern of "partnership" with the donor countries that they cannot refuse a deal proposed to them, particularly since the élites in power may derive substantial personal advantages from the different groups involved, but which are blatantly harmful to the long-term interests of their countries. Very few of them have considered the lessons arising both from the Japanese experience in the past, and from that of the present-day China. Though the performances of these two nations differ in many fundamental respects, they have two features in common: foreign capital has been, broadly speaking, kept out of both Japan and Modern China, and an elaborate policy of transfer of knowledge through acquisition of skills rather than that of ready-made technology has dominated their strategy of development.

Public opinion in LDCs is, of course, free to take a much more severe view of the post-war record of aid, both bilateral and multilateral, and to voice its mistrust of all the intentions of the rich countries. As a matter of fact, an atmosphere of gloom and disenchantment now prevails even among people who once believed in worldwide international co-operation for development.[3]

The Impact of Aid; the Growth of Distrust

(a) *The donors*

With respect to the donor countries' economy, there is a paradox inherent in the "aid game". In almost all cases, giving aid promotes one's sales thus financing out of

[3] See Mahbub ul Haq in *CERES* (FAO review), vol. 3, no. 2, April 1970, as an example of the present mood of distrust in an LDC country with respect to aid.

public money one's exports. Increasing production (in this case, for export) is always advantageous to business in a market economy, whenever idle capacities exist, as this stimulates economic activity through the well-known multiplier mechanism: GNP grows by a margin substantially higher than the initial expenditure involved. Thus, far from being a burden on the donating economy, such operations may be self-financing. Together with grossly increased outlays in social policy, they could provide a substitute to expenditure on armament in reconversion schemes. This could be carried on, e.g. in the American economy, as long as the political aspects of the problem could be taken care of.[4]

But for the donor government it is almost impossible to sell the same operation as genuine aid to the economy of the recipient country and as "self-aid" for the home economy. Hence, the impression is created that aid is a heavy burden for the donor's economy. Angry reactions on the part of the donor's public opinion are therefore to be expected because of "ingratitude" of the recipient countries and the feeling that public funds are being "wasted" on these ingrates. The good feelings arising out of the sense of fulfilled moral duty are thus often more than offset, and successful campaigns against more involvement in foreign aid get under way. The LDCs react to these campaigns; mounting mutual distrust is the end result.

(b) *The recipient country*

Inflow of foreign resources is not tantamount to increasing the LDCs capacity to invest. Quite often, external savings displace domestic, releasing them for greater conspicuous consumption of the élites. The inventory of imports financed through aid is not in itself a guarantee against this result. If essential industrial equipment is obtained through aid, foreign currency thus saved by the recipient country may be turned to importation of luxuries and to acquisition of equipment for producing luxury goods.

In first approximation, a reliable criterion for assessing the impact of foreign aid is to compare two alternative plans; one made on the assumption that no aid would be flowing in, the other on the assumption of reasonable use of a quantity of foreign aid.[5] The procedure for doing so is, however, cumbersome and difficult and in addition, it leaves out the important question of how planning decisions would be affected by expectation of foreign aid. How many responsible people would refuse aid if its conditions did not match the priorities which they have established and, as a consequence, would require changing these priorities? The currently common practice is to modify the plans so as to fit them to the programs for which assistance is available. In some cases, the whole pattern of planned development is distorted by import-bias and heavy reliance on transfer of ready-made technology, often ill-adapted to local conditions. Consciously or not, the LDC governments abdicate, in some way, their right and duty to make decisions.

[4] Hence the importance of the questions raised by John Galbraith in his recent pamphlet: "How to Control the Military", New York, 1969.

[5] M. Kalecki and I. Sachs, "Forms of foreign aid", *Social Science Informations*, March 1966.

Commonly, men of responsibility in LDCs have been educated in rich countries, and largely share the reactions and concepts of the countries in which they were educated. For many of them it might be difficult to understand what is in the best interests of their own country.

Characteristics of Genuine Aid

1. Aid to any country should contribute to its *developmental potential*, i.e. to the formation of (a) a production system able to sustain the country's expansion and/or take care of the necessary imports through growing exports, (b) an intellectual structure able to feed into the productive sector the needed novel techniques and managerial ideas, and (c) a political and administrative machinery capable of independent decision-making.

2. According to this, genuine aid is characterized by its ability to ease the bottlenecks affecting the development process, which vary from one country to another and with changes in the trend of development.

3. One of the major difficulties of successful aid is the choice of methods and priorities by the recipient country. Errors and miscalculations are almost inescapable, not only on part of the donor experts, but also on that of the LDC's decision-makers. The situation is, however, not quite symmetrical. The donors have very seldom the notion that they did something wrong. With nations, as with individuals, a common reaction to mistakes is to accuse others. In the case of the recipient country, error and sufferings caused by them may contribute to education. This is why proper aid procedures should include a component of self-education by the recipient country, and require giving it an entirely free choice of the priorities.

4. We accordingly propose that the choice of forms and specific items of aid should be left entirely to the recipient country. The financial limits within which this choice can be exerted should be established in each case, for a limited period of time, after which the aid program would be renegotiated for a subsequent period.

5. Aid should be granted on the understanding that the recipient country will endeavour to use it to release internal reserves and capabilities; but the recipient country should be given the chance of learning by bona-fide mistakes. The ability for decision-making can be improved only through a learning-by-doing process, and the cost of such apprenticeship reflects, to some extent, the educational gap. In consequence, the recipient country should be credited with a certain sum; it would freely set, within the limits of this sum, its shopping list of goods, facilities, technical assistance and expert consultations.

6. The country's performance, assessed by the government interested and by an independent expert committee, should be taken into consideration as one of the criteria for periodical renegotiation of the amount of aid at the end of each period.

7. The duration of a period will be of 5 to 7 years with evaluation of the performance at the midpoint of this period and the chance to negotiate the renewal of the aid contract at that time.

Depersonalized Aid

We think that in order to free aid from second thoughts and calculations on both parts, it should become "depersonalized". This aim could be reached by giving absolute preference to multilateral aid channels. It is of the greatest importance that the U.N. aid machinery should be modified, making it less bureaucratized, more flexible and more imaginative.

The U.N. should provide the funds and keep at the disposal of the LDCs all the necessary information and advising services, thus becoming a clearing house for all data relevant to development (and all needed computation devices). Such an operation would require finances of several times larger than what is made available at the present in the United States. Hence the importance of endowing the U.N. with new and independent resources of income, arising from the utilization of resources belonging to the whole mankind. Their definition would deserve very elaborate studies (to start with, they include the submarine resources, as pointed out at the Pacem in Maribus conference held in Malta). To the extent to which this could be achieved, the very notion of donor and recipient country would fade out.

We realize that the proposals sketched above are at the opposite pole from the current practice, and that, in the short run, they may sound quite unrealistic. However, a bold departure from the routine evolved in the last quarter of century is necessary.

Cooperation in Science and Technology (ST)

The above described approach could be tried within a relatively restricted field—that of assistance to LDCs in the realm of Science and Technology.

There are three general reasons to start with ST:

1. It is necessary to have some testing grounds for the new approach. A pilot project is needed before the method is adopted on a wider scale.
2. Aid in ST can be isolated from a broader context because of special aspects of the problems involved, and of less powerful emotional and political components.
3. With respect to ST, the situation in LDCs is alarming. ST alone cannot redeem LDCs. It is not a substitute for social change or the only agent for such changes. Conversely, ST cannot fully develop without a suitable political and social environment that would give to scientists minimum guarantees of stability and of freedom of research. But, with all these qualifications, ST constitutes a critical factor of progess, being a vital part of the developmental process.

ST Colonialism

The following quotation from Calder's latest book reflects accurately the present state of things. "No caricature is involved in describing modern science as a European invention which enabled the white nations to achieve military, economic and cultural domination over the rest of the world, and to make themselves prosperous while leaving the natives of the poor countries to progress very much more slowly. No injustice is done, to say that most research workers and technologists have unthinkingly connived in these uses of science which are, at bottom, racist. Declarations about using science to feed the world's hungry have not stopped the prosperity gap growing wider; nor can they alter the fact that the intellectual interests of the great majority of research workers are far removed from any such program, and that the preoccupation of technologists is with machines that enrich the rich."[6]

As a matter of fact, according to calculations made at the Science Policy Unit of the Sussex University, roughly 98% of the non-communist world expenditure on Research and Development (R & D) is confined to the U.S. and Europe. Thus two-thirds of (non-communist) mankind accounts for only 2% of R & D expenditure, the discrepancy in per capita ratio being no less than 100:1.[7] As the advanced countries have a virtual monopoly of R & D, the selection of priorities of research and of methods for solving the problems takes into consideration the needs, resources and the capacities of these countries, while in most cases the interests and the situation in the LDCs are pointing in completely different directions. Moreover, much of the present expenditures of the poorer countries represents a hopeless attempt to compete, from an inferior position, in solving the same kind of problems, by the same methods, rather than to attack problems suggested by their own conditions. In fact, the indigenous scientific and technological capacity of the poorer countries is even insufficient to determine the nature of their own problems and to determine how far they are susceptible to solution by applied science and technology with appropriate methods.[8]

Indiscriminate transfer of ready-made and ill-adapted technology to build pockets of modern industry in LDCs cannot solve the basic problems of underdevelopment. Although some LDCs can proudly show now some booming industries and modern cities, social inequalities and unemployment may have worsened. According to H. W. Singer's estimates, the present level of unemployment in LDCs is at least 25%. Moreover, it has been found that in fourteen LDCs, unemployment increases at a rate of 8.5% per annum.[9] On the assumption that the present trend continues unaltered, one is led to estimate that in 1980 unemployment in LDCs will grow to 43%.

Some Proposals for ST aid to LDCs

Two kinds of solution are being proposed:

1. Scientists and technologists in the developed countries should pay more attention than hitherto to specific problems of the LDCs. Some 5% of expenditure on "civilian" R & D in developed countries should be devoted to such purposes.

[6] Nigel Calder, *Technopolis—Social Control of the Uses of Sciences*, London, 1970, p. 252.

[7] Data quoted by Hans W. Singer in *Dualism Revisited*, October 1969, issued by the Institute of Development Studies at the University of Sussex.

[8] Hans W. Singer, *loc. cit.*

[9] Calculations of H. A. Turner quoted by H. W. Singer.

2. Effort should be directed at expanding the scientific and technological infrastructure in the LDCs.

If properly implemented, 1 and 2 should be complementary. But the present pattern of ST domination could easily find ways also into a program of large-scale research on behalf of and for LDCs by scientists and technologists in advanced countries. The results transferred in ready-made form could even worsen the present situation. The resources potentially available should be therefore put at the disposal of LDCs, with the expectation that they would use them, by free choice, to promote their STI (by STI we mean an integrated system of scientific and technological information services, research institutions—working both in the field of fundamental and that of applied research—projects, designing and engineering offices, and agencies entrusted with the control of international transactions in the realm of technology).

The activities of the U.N., in the field of ST, should be in the same spirit as described above. An allocation of resources for, say, 6 years subject to renegotiation, should be made in terms of a percentage of the R & D budget in an LDC. It is anticipated that this percentage will decrease with time, being adjusted in the light of its performance evaluated by an independent agency, say, every 3 years by the LDC; this agency should be jointly appointed by the government concerned and a committee of international experts selected by the U.N.

The recipient country should be entirely free to spend the resources allocated according to its own priorities in research, using the best available sources of expertise and hardware. Only one limitation would be imposed: research made possible by the aid should be exclusively of a civilian nature; no resources could be diverted to military projects.

The U.N. would provide to the LDC a number of back-stopping and ancillary services such as:

Data banks stored up with up-to-date scientific and technological information, including the "shopping opportunities" and the rosters of experts.

Facilities for quick trips and consultations abroad, made available for scientists and technologists working in the LDCs, who should be free to pay short and frequent visits to their counterparts in any country. IATA's regulations should be amended so as to allow for the use of unsold seats in airplanes for a nominal fare.

International research institutions, preferentially located in LDCs intended above all as meeting-places, and endowed with training and educational capabilities. Such institutions, together with a certain number of their counterparts in social sciences, could eventually be federated into an International Development University. The main vocation of this University would be fostering of interdisciplinary development-oriented research and training at the highest postgraduate level.

[10] A proposal for setting under the U.N. auspices of an International University was being discussed in the United Nations at the time of writing of this note.

The proposals sketched in this note are very tentative; they aim at drawing the attention to the need of a radical departure from the existing philosophy and machinery for international aid. Both the Pearson Report and the Jackson Report are, in our opinion, much too conservative in this respect.

Social Sciences and Social Development[*]

S. Dedijer†

"It is not necessarily difficult to know,
but it is certainly difficult to act."
Wang Mang Min, 1472–1528

"Action is easy, understanding difficult."
Sun Yat Sen, 1866–1925

The demand for a scientific approach to the aid of the developing countries is one manifestation of a new and hitherto little recognized feature of what is commonly called the scientific or scientific-technological revolution. In this new phase of the revolution all the world's major problems are becoming problems of basic and applied social and behavioural sciences, of formulation and implementation of scientifically based social policies. Peace, population control, food, disparity of development and other major aspects of the process of formation of a world culture are now being recognized and treated as social and behavioural sciences problems. The major problems and policies of every country within this process, irrespective of its size, degree of development or political system, cannot be tackled any longer simply through the intuitively summed up experience and "know-how" of practical politicians, administrators, strategists, industrialists, educationists, or even natural scientists, etc.; it must be backed by a scientific "why" of these problems and of the actions they require, supplied by the social and behavioural sciences. The apparent contradiction between the two bits of Chinese wisdom cited under the title, is now being resolved in that the men of action must base their policies on knowledge and understanding supplied by the social sciences research. Each particular social system can effectively develop today, within the world system, only with the help of a continuous social science reappraisal within the constraint of the world environment, of its own values and objectives, of its own social evolution, of its own mechanism of social heredity, and of its own current policies for social and cultural change and development. The effort within a large number of countries to identify, to attach these problems and to formulate rational policies in economics, education, defence, science, foreign policy—including the policy of aid to the developing countries—is giving rise to a growing, through unequally distributed, world demand for the

* Presented at the Pugwash Conference on Science and World Affairs, Addis Ababa, Ethiopia, 1965.
† Research Policy Program, University of Lund, Lund, Sweden.

development of research work in the social and behavioural sciences. The rise of this new feature of the scientific revolution is shown by the fact that in all the developed countries, irrespective of their political system, there is, as it has been pointed out, a growing awareness that the development of a social sciences research policy is equally important as the development of policies for the basic and applied natural sciences. Most developed countries, regardless of their size and political system, are now energetically reviewing their own social sciences policies.

Underdeveloped countries—and let me here, very inadequately, define them as those with a national income per head of less than $200—are still on the whole passive by-standers in the scientific and technological revolution. I suggest that in this we shall find one major cause—but by no means the only one—for the growing gap in development between them and the rest of the world. In these countries, research work, basic and applied in natural and social sciences, is still a nonexistent or an insignificant category in their social division of labour, whether we measure it in manpower, expenditure, research results, social demand for research, or in terms of organizations set up for its development. Furthermore, I conjecture that the undeveloped countries develop social sciences slower than their own basic and applied natural sciences. Hence, in my opinion, the most important aid that can be given to these countries is the extremely difficult problem of helping them to develop their own indigenous social and behavioural sciences. To approach scientifically the problem of aid to the developing countries it is necessary to stimulate a corresponding scientific approach on the part of these countries themselves to their own problems of development; to stimulate a corresponding effort to base their own policies, including the aid seeking and receiving policy, on their own indigenous combination of practical "know-why".

To begin to deal adequately with the question of social sciences policy in the undeveloped countries one should consider systematically:

1. why social science should be given a high priority in the less developed countries;
2. what factors contribute to the low demand and slow development of their social sciences;
3. how should social sciences be developed there; and
4. how should their results be used to guide policy actions.

Here I shall briefly touch only on a couple of major aspects of this problem. And to start, I shall consider briefly the changes in social science theory of development. The great contribution of Karl Marx, whom Julian Huxley calls the Bacon of social sciences, was that he showed that to understand and to engineer social change it is necessary to understand and to change the productive forces of a society. During the past generation, when the problem of disparity of development became a world problem, almost all the aid policies of the developed countries—even of those officially "anti-Marxist"—and most of the social science thinking both East and West on this subject, have been based on the theory that the development of the productive forces in the less-developed countries is not only the necessary but also the sufficient condition for their social development. The failure of aid policies of many countries to achieve their main avowed objective—a rapid social development of the countries they aid—has been a contributing factor to revision not only of these policies but also of the theoretical thinking on the problem. Today economists, both West and East, are introducing in their production development equations not only the classical economic variables—such as labour and capital—but also such "residual" factors as scientific research, education, management and administrative practices, socio-psychological factors, etc. Social sciences are coming to the conclusion that economic development is not possible without changes in all the basic elements of a culture: including institutions, social structure, social patterns of behaviour such as sex, child rearing, family structure, language, religious practices, etc. In other words, social development must be seen as a problem in a planned, conscious, simultaneous change of all aspects of a country's culture. As a prospectus for the United Nations Research Institute for Social Development, published in Geneva in September 1964, says: "It has thus become clear that a purely sectoral approach to development policy and development planning is inadequate and that an integrated approach is necessary."

I would like to suggest that the necessary and sufficient conditions for relatively rapid social development can be considered to be as follows:

(a) *National ideology*, including the ideas, values, objectives of development; norms of behaviour; and the *will to develop* through independent national effort shared by a significant fraction of a country's population,
(b) *Organizations*, political and social, translating this ideology of development into a pattern of
(c) *Actions*, for the formulation, planning, execution, of social policies, all involving not only the decision-making elite, but also the broad sections of the population; all three of the above factors being based upon a
(d) *Social communication system*, continuously and creatively producing, processing, transmitting, public, objective, scientific information free of dogmatism and doctrinairism put out by the indigenous social and natural sciences research:

(i) about the outside world environment and its interaction with the political, economic, scientific and other aspects of the country's life and culture;
(ii) about the productive forces of the country, including the natural resources, and the indigenous research potential in natural and social sciences and the problems of development of all of them;
(iii) about the local conditions: scientific, including information on the country's social evolution, and structure, its mechanism of social heredity and social change and on general patterns of social behaviour in relation to national needs and goals;
(iv) about the changes in national ideology, organizations, actions which are necessary in order to pursue and achieve the national objectives in a stably evolving yet creative political and social environment.

From this, three things follow:

First, none of the major problems of the developing countries such as the provision of adequate food supplies, improvement of health, a more complete understanding of population questions, the most effective use of the natural resources, the problem of industrialization, the invention and application of new educational techniques, etc., can be treated adequately purely as a question of basic or technological research in natural sciences or even in economic science. They must be considered primarily as problems of social and cultural change and hence they must be tackled by the multidisciplinary approach of both natural and social sciences.

Second, a scientific analysis and planning of economic and social development must include a scientific analysis of the politically necessary conditions for that development.

Third, the existence of the politically necessary conditions—expressed in national independence and in political organizations uniting and channelling the national will to develop—is not sufficient for rapid social development. The collective will to change social conditions, to develop, is a most valuable resource for progressive social change—but it is only a raw resource which can be wasted and dispersed as ineffectively as any other resource. The will to change a society, to develop it—expressed in political organizations and actions—can be effective only if combined with a continuously growing, increasingly automatic social communication system, producing objective, public information by means of the social and natural sciences. Social sciences, objective and public, free of doctrinaire ideological pressure, and of immediate short-term political constraints, must supply the social engineering knowledge for effective growth and for the effective use of the most precious of development resources: the national will to develop through self-reliant effort. And here we are faced with a difficult problem in social theory and even more difficult in social and political practice. A scientific approach to the problem of aid to the developing countries, if it is to be effective, must not avoid to face for political or ideological reasons this particular difficult problem of development. The problem consists in this:

The organic unity of the present world is expressed among others by the fact that every country, irrespective of its size, degree of development and political system, is exposed to pressures of the world environment. The less-developed countries, the small ones among them to a highest degree, are exposed to these pressures, economic, political, social, cultural, very often in the form of policy actions of the largest countries. The release of the will to develop, its maintenance and the maintenance of the very national independence, call for the recognition, identification, scientific knowledge of these outside pressures and for realistic policies to cope with them, to co-operate or to resist them. At the same time, the development of a social communication system producing objective, scientific information necessary for development calls for the promotion of social and natural sciences obeying their own rules of existence and development. Social sciences have the task to present continuously a critical image of the main problems and policies of the country and to produce objective advice on their necessary changes. It is now generally accepted in the developed countries, irrespective of their political system, that all sciences, including the social ones, can develop and function effectively only if they develop under the conditions of freedom of criticism, of struggle of various opinions, schools of thoughts untrammelled in their search of problems, methods and in publication of their results by political and other institutions.

In real life practice up to today, the leaders of undeveloped countries, irrespective of their size or political system, have always considered that the full development of a social communication system based on social and natural sciences, operating on their own inner rules (as such a system exists in almost all the developed countries) is irreconcilable with the maintenance and growth of the national will to develop and the maintenance of national independence. Other important factors than this alleged contradiction tend to give a lower priority to the development of social science in the less developed countries. Such are, for example, the ideological doctrinairism, the lack of information about the new practical role of social sciences in the developed countries, etc. But the main reason, in my opinion, for the lack of development of social sciences of a modern kind in a number of less-developed countries lies in this belief in the alleged contradiction between the development of social sciences and the maintenance of the national unity and will to develop. It is a major challenge to the developed countries, and to their scientists in the first place, to keep showing in an objective way on the basis, among other of their own social practice, that these two essential elements for effective social growth and development are not exclusive of each other.

*Science for Development—What Individual U.S. Scientists Can Do**

CHARLES WEISS, Jr.†

ONE of the most important tasks for the American scientific community in the next 5–10 years is to help organize groups of scientists, both here and in the developing countries, who are concerned with the problems of development. This challenge calls both for national leadership and for grass-roots efforts: well-funded research institutes, grant programs, journals and organizations; and modestly funded programs for small groups of individual scientists asking themselves, "what can my talents and the resources under my control do to further the contributions of science to development?"

This assertion may surprise scientists who are accustomed to thinking of development as primarily an economic and political problem. But the widely publicized ecological misfortunes caused by the Aswan Dam are only the most recent manifestation of the difficulties caused by insufficient appreciation of the technical aspects of development. On the brighter side, the "Green Revolution" from the new cereal grains has proved the value of technological research by bringing a number of countries which were previously net importers of food to the brink of self-sufficiency.

Science for development can be defined as including those aspects of natural and allied social sciences, including technology assessment, that relate to the problems of LDC's. This is a broad definition that includes both basic and applied topics. I do not think we need to be fussy about the distinction, as long as it is clearly understood that the subjects of fundamental investigation are to be chosen for the short- or long-term relevance to development.

The problem of launching a new, multidisciplinary area of research is not new to the recent history of American science. Oceanography and biophysics were in similar positions 20 years or so ago. The basic tasks are to bring to the attention of the American scientific community the existence of exciting and important problems, to make available research money and to publicize that availability, and to provide the communications necessary for the establishment of a cohesive new discipline, that of Science for Development. In this way, we hope, a climate of opinion will be

* Presented at a meeting organized by the U.S. Agency for International Development in Washington, D.C. on March 4, 1971.

† Research Staff Member, IBM. T.-J. Watson Research Center, Yorktown Heights, N.Y. Presently Science and Technology Advisor, International Bank for Reconstruction and Development, Washington, D.C.

created within individual departments and laboratories regarding research on problems related to underdeveloped countries as a valuable discipline worthy of full-scale professional recognition.

Once such a climate of opinion has been built up, the American scientific community can become an extraordinary resource to developing countries. Here we have an immense reservoir of talented and vigorous men and women who are asking themselves what they can do, eager for the kind of challenge that development problems present.

Many important contributions to the technological problems of development can be achieved by taking advantage of the skill, imagination, energy, and goodwill of scientists individually or in small groups working in the United States on relatively modest budgets. Some of these contributions actually amount to changes in the attitudes of university faculty as expressed in curriculum planning, academic policy and appointments. These potentially far-reaching developments cost practically nothing. Others are "mini-foreign aid programs" that can be run by a small staff on a modest budget, in Washington or at a university.

This paper will present suggestions that arose from numerous discussions and meetings with scientists with varied experiences in less developed countries. I feel that there is now a fund of ideas more or less common to all scientists concerned with development problems that recur with variations in virtually all articles and lectures on the subject. There is no shortage of good ideas. It is time for a frankly experimental approach.

The most obvious way for individual scientists to help the less-developed world is by doing what they are best at, research. The list of possible research topics shows the terrible ignorance that covers even some of the most fundamental aspects of the tropical environment in which most less-developed countries (LDCs) exist. These research problems are by no means academic. A large fraction of the foodstuffs grown in LDCs is eaten by insects, rodents and worms, many of whose habits have not been thoroughly studied. The properties and ecology of tropical soils are imperfectly understood, so that much agricultural extension work takes place in a scientific vacuum. Science education is primitive in many LDCs, so that citizens from peasants to cabinet ministers are unprepared to appreciate vital technical issues.

A partial list of relevant research topics follows:

mathematical modelling of tropical ecologies
implements for the family farm
water management
ecological consequences of development
(one-crop farming, dams, fertilizers, insecticides, clearing of forests)
population control
pandemics
technology transfer to countries with excess labor supply

social consequences of technological change
tropical meteorology
tropical oceans, seas and lakes
taxonomy of local plants and animals
low-cost village or slum water supply and sanitation
simple machinery for processing agricultural commodities
new uses for natural fibers.
farming techniques for unirrigated tropical land.

There is plenty of room for both "basic" and "applied" research in these areas. In fact, the distinction between basic and applied work is often fuzzy: taxonomic classification of local plants and animals is basic to a biologist, applied to an agricultural or fisheries scientist. Such research could well be carried out in the U.S. in collaboration with scientists working in LDCs. I hasten to add that social scientists could easily prepare long lists of their own.

A related area worthy of special mention is the development of primary and secondary school curricula for science teaching in less-developed countries. Here I want to cite specially the wonderful work of the Education Development Corporation on primary science and mathematics teaching that I saw 7 years ago in Africa. This work was done in the field on modest budgets by small staffs. For all that, I consider it some of the most intellectually exciting research on the whole continent.

A Science for Development Curriculum

A second major area for contributions by United States scientists lies right in their own backyard: in the structure of the university curriculum. I would propose that American universities should and will soon offer undergraduate and graduate programs in Science for Development, programs which will combine basic natural sciences with social science, regional studies and perhaps public or business administration. I would hope that this will not be a watered-down science degree but one tailored to the requirements of a new breed, the well-rounded development scientist. With the new discipline should come a career pattern which will alternate university teaching and research with technical assistance abroad and perhaps government service.

In building such a discipline, we start with one tremendous advantage. On American campuses today is an idealistic new generation of students that is eager to experiment with a new and highly relevant area. I think students are always a little ahead of most of their elders in sensing this kind of trend. In any case, discipline-oriented science is suffering from a severe soul-searching and financial buffeting.

The new spirit on the campus gives us a chance to face the classical "brain-drain" problem in a new and different way. One of the major contributors to the brain-drain has been the feeling among scientists from less-developed countries that the only problems worth doing are those on which is focused the attention of the international scientific community, and that the only place that research can be done is in high-powered laboratories with the latest equipment. To a large extent, these feelings have been a reflection of the attitudes of the faculty and students in universities in developed countries. Visiting LDC students cannot help but be infected by the new atmosphere on campus. We must prepare them to face the challenges of doing science in a less hospitable environment, and reinforce their determination by encouraging the attitude that Science for Development encompasses some of the most pressing and important problems in the world today.

An introductory course on the technology of development should be aimed at making students aware of the potential importance of research and technology to development problems. Perhaps in the long run such courses will help increase the willingness of LDC governments to fund technical research, and thus provide financial support to keep their scientists at home and productive.

Science students from LDCs are generally given the same preparation as other students, despite the very different conditions and problems they will have to face when they return home. The situation differs greatly from university to university and even from student to student, suggesting that important contributions can be made by concerned individual faculty members with or without guidance from larger organizations. One approach to the special problems of the LDC science student in America is through the Science for Development curriculum proposed earlier. A simpler method is for individual professors and departments to grant their LDC graduate students leeway in choosing courses and seminars, even outside the usual departmental limits, which will prepare them for the unique conditions they are likely to encounter at home.

Science students from different LDCs often do not realize that they will face many of the same problems at home, no matter what part of the world they come from. They will be isolated from fellow scientists, far from suppliers and repairers of essential materials and equipment, and responsible for administrative matters relatively early in their careers. They will be members of small scientific communities in countries where science carries little prestige—yet almost the only nationals able to give competent scientific advice when it is needed.

I think that every university with science students from LDCs should sponsor a Development Science Club or seminar which will enable these students to get together and talk about the problems they share. A modest budget would enable them to invite outside speakers and to do small research papers, but even this is not absolutely necessary. These clubs are clearly within the capacity of the individual interested scientist—especially if the department will count it against his teaching load.

A second need is for a summer course lasting 3–4 weeks to introduce LDC scientists to the art of running a laboratory. This is a larger undertaking but still qualifies as low budget, small staff. LDCs lack trained administrative staffs, and new Ph.D.s there are likely to be faced quickly with such tasks as setting up a library, a machine shop, or even a whole new building. They need to know something about the structure of the scientific community and its relationship to government and society in different parts of the world. A summer institute would enable the LDC student to concentrate on science during the year while still getting essential preparation for his return home. (American students and senior scientists might benefit from such a summer course, but that is another matter.)

In general, every college and university can examine its own programme and ask whether it cannot, with relatively small and inexpensive changes, be adapted to the needs of development. Teaching colleges, nursing and medical schools—almost any institute offering technical training—can institute courses or clubs to focus on the special problems of development. Indeed, I hope that student interest will be high enough that these will be instituted by popular request.

Individual U.S. scientists are likely to face a further question. Should an LDC science student choose a thesis topic directly related to his needs on his return home? A variety of opinions exist on this subject, and the answer no doubt varies from student to student. The important point is that the question be faced frankly by the student, the professor and the department, and that real options exist. If the student and his professor feel that any good research is valid intellectual preparation for research on any subject anywhere, I doubt that anyone could muster convincing arguments to the contrary. On the other hand, there are advantages to choosing a problem which is relevant to the needs of the home country (and perhaps chosen in consultation with scientists there) but best adapted to being studied in the U.S. While pursuing his research, the student might well be alert to eventual simplifications in technique that will enable him to continue the work at home.

An ideal arrangement for many students will be to do graduate courses and general examinations in the U.S., choose a topic which can be studied in the home country in association with the university there, and then return to analyze and write up his findings in the U.S. publication in a form accessible to the home country should be encouraged or required. (This is a frequent problem with visiting social scientists.) This sequence offers the student a first-hand orientation to the research environment at home, as well as useful contacts with the important scientists there. It also facilitates job placement. Unfortunately, no fellowship program exists at the moment to make this arrangement possible.

Travel Funds: An Urgent Need

The greatest problem faced by scientists in LDCs is isolation. Individual scientists and departments can contribute here by giving serious consideration to top-quality junior scientists from LDCs in filling visiting lectureships and professorships under their own authority. Two or three such people might be available for the price of one "big name"—and their visits might well be more valuable to the U.S. department for the new perspective they bring to it.

The major need here is for travel funds, travelling fellowships, and journals and other written instruments of communication. Some of these can be run by small staffs operating out of universities; others will list some of the pressing needs and hope that others present will be able to suggest the proper institutional framework.

1. Fellowships for thesis or postdoctoral research within LDCs by students at American universities. These students could be Americans or LDC nationals.
2. Secondments for U.S. scientists to enable them to spend 2–4 years in LDCs and return to their old jobs with due recognition for the service they have performed and the experience they have had. This system has been in effect for years in England and France, and has been a main conduit for staffing LDC universities with Europeans who will for some time be urgently needed.
3. Fellowships for sabbatical travel and research by LDC scientists. Visits to other LDCs should be encouraged, not just to the United States.
4. Fellowships for training in special topics, such as computers or R & D management.
5. Journal(s) of science for development. One function of such a journal would be to aid communication between the LDC (where the problems are) and the developed countries (where students and equipment are more plentiful).

If all goes well, Science for Development will become a sprawling field, and there will be a great need for long-term perspective on the one hand, and for handbooks of "packed down" information for use by policy making officials and planners on the other. Both of these are best provided by individual scientists, although the National Academy of Sciences and other organizations may well wish to play a role. An example of the latter type of material is the "Early Warning Handbook" prepared for planners of large-scale projects by the Food and Agricultural Organization of the United Nations. Another is the book by Professor McKee of Caltech, *One Hundred Problems in Environmental Health* (Jones Composition Co. and Kirby Lithograph Co., Washington, 1961) which contains a brief statement and basic bibliography for scientists looking for an interesting and important problem in this area to tackle.

I want to conclude on a more philosophical note. Dr. Carlos Varsavsky of Argentina has developed the concept of the "inward-looking scientist" in an LDC, who, whatever the subject of his own research, is concerned with the problems of his own country and trains his students to share this concern. The inward-looking scientist is not necessarily doing applied or even mission-oriented research, although it is reasonable to suppose that he would accept such a problem if it were important to his country and appropriate to his training. The implication of Varsavsky's argument is that there are in LDCs "outward-looking" scientists who work on the problems currently fashionable in international science and look at their own country only as a source of funding and of students.

The importance of this argument is not to attack "outward-looking" scientists—many of whom are making important contributions—but to emphasize that the primary purpose of the scientific community in an LDC is to effect technology transfer and thereby social change. I am certainly not arguing that all research in LDCs should be tied to immediate industrial application—far from it. But I do not think that we can consider the building up a community of basic scientists as an isolated problem and assume that the technology transfer will take care of itself. On the contrary, we must be thinking about technology transfer from the beginning. Those who wish to help science in LDCs should keep in touch with the applied scientists of such technical industry as exists in the LDC and ask them at every stage what *they* need from their country's scientific community. In many cases, they will not know themselves—which raises a new problem. What should the interface between industry and university science look like in an LDC? Unfortunately, scientists in DCs familiar with the problems of LDCs are generally not familiar with the field of technology transfer—yet these are the scientists who will train scientists to return to their home countries and face these very problems.

Perhaps it is impractical to dream of the widespread popular interest in development that the above discussion implies. After all, the number of scientists now interested in development is small compared to the magnitude of the problem. But even 5 years ago interest in ecology was limited to a small band of devotees. What subject affects our environment more than the future of two billion human beings and their descendants?

*The Role of Technologic Institutes in Industrial Development**

LAWRENCE W. BASS†

TRANSFER of scientific knowledge to practical use is dependent on the closeness of interaction between technical experts and industrial managers. This is a basic lesson from the commercial history of the past century. In developing countries technologic institutes can be effective agencies for this transfer.

Science-based industry in the United States required several decades to progress from dependence on foreign research to its present strong position. A similar history is found in other advanced countries which had formerly been backward in developing the complex of skills and institutions necessary to become competitive in technology. Each of these nations evolved its own mechanisms to stimulate technical entrepreneurship.

Developing countries are striving to short-cut such prolonged gestation for internal competence in technology. In this effort their perception is clouded, however, by failure to understand the difference between scientific research and the practical measures needed to introduce its fruits to enhance industrial growth. Mere creation of scientific knowledge is not enough: economic development will take place only to the extent that industrial managers can be convinced of the benefit of using this specialized know-how.

Major new industrial undertakings in developing countries—and often also in advanced countries—are now and will continue to be based on imported technology. The foreign expertise has the vital advantages of greater confidence in reliability, faster implementation, and lower risk in investment.

Smaller firms in these countries have much less access to imported technology, yet collectively they account for a large share of national production and employment. Hence any program to make improved use of local technical skills should nurture their help to medium and small enterprises.

Why Companies Seek Outside Help

As the basis for discussion of the role of technologic institutions, one should first analyze the reasons for their existence. Large and sophisticated companies in advanced countries are frequent clients of contract technical organizations, notwithstanding

* This paper first published in *World Development*, Volume 1, Number 10, October 1973.

† Consultant, formerly Vice President Arthur D. Little, Inc., Washington, D.C.

their internal strengths of expertise. Their motivation for these external services, as summarized below, holds true likewise, not only for smaller enterprises, but also for developing countries.

1. *Obtaining skills outside their own resources* in preference to the slow and costly route of developing and maintaining in-house expertise until its utility has been proved.
2. *Handling work overloads*, because companies cannot afford to support staff to take care of steep peaks in demand.
3. *Unbiased analysis of internal findings* to reduce the risk of embarking on new undertakings.
4. *Settling of internal differences of opinion* through objective evaluation to curtail protracted arguments within the organization.
5. *Stimulation of in-house skills* by acquiring new methodology from outside.
6. *Securing disinterested appraisal of projects* to back proposals for financing.
7. *Support for public relations* to confirm claims for the merits of products, with due regard to the ethnics of institutional endorsements.
8. *Obtaining unbiased intermediation* in dealings between independent organizations.

Granting the cogency of these reasons, and re-emphasizing the need for close interaction between supplier and user of expertise, we can frame an axiom of successful consultancy: both consultant and client should observe the letter and spirit of the concept that the former serves, for the purpose of the project, as an integral extension of the latter's organization and is whole-heartedly devoted to the client's interests. Hence the mutual obligations must be clearly defined in a contract, and significant deviations from it must be reviewed to ensure continuing agreement.

Channels for Technical Help to Industry

An obvious path to improve technology is to encourage the support of internal technical programs in individual firms. This is a slow process, but it can be speeded up by government through fiscal incentives and by professional and industrial groups through educational and promotional efforts. The effect, however, will be largely confined to larger firms because small units are rarely in a position to sustain such activities.

Government-applied research laboratories have been founded in most developed and developing countries. While many of them do very good work, they are usually restricted by policy from dealing with the questions of individual private firms and therefore their programs tend to be directed toward investigations of general problems of sectors of industry; then, too, confidential treatment of information, which is essential for productive relationships, is difficult in public institutions. Their activities are hampered by bureaucratic procedures and by lack of employment opportunities to attract the cream of the technical community. Few countries are complacent about the ability of their government laboratories to instill by themselves the impetus of new technology needed for industrial growth.

Primary responsibility for technical help to industry is not usually placed directly on the universities because their dominant functions are education and scientific research. Nevertheless, many faculty members, because of their practical interests and skills, have participated in the introduction of improved technology. While such activities depend on the happenstance of collaboration between a technical specialist and an industrialist with a problem, the probability is increased by a coordinating internal focus to provide a channel for these services. Technologic institutes have grown in a number of cases from recognition of this function.

Quasi-public laboratories to serve selected sectors of industry were established, with government backing, in several countries in Western Europe during the early decades of this century. They were usually founded through the medium of an industrial association and they have been subsidized in part by the public purse. They have carried out studies on general problems and have helped to spread the gospel of the merits of applied research. Their scheme of operation is not well adapted to serve the specific needs of individual firms, which must instead assume responsibility for selecting and using the technical information which they provide.

In the United States a different type of institution has come to the fore, namely, the contract research organizations. These present the great advantage of face-to-face contact between the supplier and the user of technology on specific problems. Private consultancies for this purpose were established about a century ago, and in 1907 the idea of a multi-purpose industrial research institute took root under the aegis of a university. This concept found increasing acceptance between the two World Wars and has expanded vigorously in the last three decades. The scheme spread to Western Europe 20 years ago and there the rate of growth of such organizations now exceeds considerably that of the industrial association laboratories mentioned above.

These media for promoting technical advances are strengthened by the efforts of suppliers of materials and equipment to expand their markets. While they naturally promote their own products, competition among them offers purchasers a range of options in making their selections.

Technical Services in Developing Countries

All these measures are being used in varying degrees with varying success in developing countries. There has been great stress on increasing the scientific capabilities of the universities. These, however, as pointed out earlier, do not automatically become mobilized for constructive attack on the problems of industry. Too often there is little interest in the routine aspects of technology on the part of faculty members because of their leaning toward prestigious scientific research. Success is more likely when a university installs what is in essence an internal technologic institute with a coordinating function to encourage a flow of inquiries that can be directed to the appropriate technical specialists. In expanding educational programs there is unfortunately inadequate attention paid to supply–demand factors to induce more effective use of technical manpower.

Quite a number of technologic institutes on the American model have been set up in developing countries, often with initial support from USAID or other international development agencies. Many of them are called industrial research institutes, but in Latin America the name institutes of industrial technology is common; in my opinion this is a better title to point out their primary mission.

The performance of these organizations has been spotty and only in rare cases have they lived up to the optimistic expectations of their founders by becoming dominant factors in industrial development. The chief criticisms in local circles are that they have not been the source of major new industrial enterprises, that they have not won extensive acceptance by industry, that the services they perform are routine rather than innovative research, and that they have not recruited highly trained scientists.

These complaints are, in my opinion, only partly valid. They rest on misunderstanding of the facts of life about the industrial complex in a developing country. This lack of understanding pervades government agencies, the industrial community, academic circles, and even the institutes themselves. It is the purpose of this paper to present a pragmatic view of the ways in which these institutes can improve their programs to meet the real needs of their national economies.

In spite of initial misconceptions and handicaps, a number of technologic institutes have shown creditable performance, as illustrated below by specific examples with which I am personally familiar. They have taken advantage of their assets of diversified technical skills to create a place for themselves as important components of national technical resources. From these case histories it is possible to outline the directions in which such organizations should develop to attain optimum potential.

Japan, with defined national objectives for industrial development, has benefited from a decentralized complex of over seventy technical centers and over fifty management centers to assist industry, particularly the smaller firms. These are supported chiefly by the prefectures and large cities, with overall coordination and modest financial help from the national Small Enterprise Agency. The system in principle resembles that of the state agricultural experiment stations in the United States, except that the latter bring to bear the widespread impact of county agents to interpret technical information to the individual farmer. The Japanese centers concentrate on the technical and managerial problems of small industry in their particular localities, thus bringing the sources and users of knowledge into closer proximity. They have been especially helpful in upgrading the quality of manufactures to meet export specifications as well as requirements of other firms which they serve as suppliers. These two parallel programs of technical and managerial assistance would benefit from closer interaction.

Singapore has established within a decade an integrated complex of capabilities to aid the community in developing industry to supplement its historic function as an entrepôt. The coordinated services range from analysis and testing to product and process improvement and development, industrial design, export promotion, management practices, and techno-economic evaluation to support applications for development loans. This model deserves adaptation in other countries.

Ability of a technologic institute to obtain a major part of its operating funds from charges to clients is a good measure of its acceptance by the industrial community. In the United States and Western Europe these organizations are generally self-supporting, but this happy situation cannot be reached in developing countries until a vigorous base of industry has been attained.

Of the twenty-eight laboratories in developing countries which reported contract fees in the 1972 Directory of the World Association of Industrial and Technological Research Organizations (WAITRO), about one-half showed earnings of less than 25% of expenses, one-quarter in the range 25–50%, and one-quarter over 50%. These data cannot be compared rigorously because of differences in computing operating costs and variations in defining fees from contracts. Nevertheless, the figures are encouraging by confirming the fact that some institutes are quite successful in attracting clients.

As an example of progress, IPT in Brazil in 1969 had earned income of only 20% of expenses, reached 40% in 1972, and hopes for 60% in a few years. The Industry Institute in Beirut has attained a high degree of self-support by extending its sphere of activity outside Lebanon, from which about half its income is derived, to other countries in the Middle East which do not have adequate technical capabilities. IIT in Colombia is an interesting case of an organization which, instead of at once spending funds and effort on an elaborate building, got along in adapted preliminary quarters until it could justify the acquisition of new facilities.

Technical and managerial services for small industry are being pioneered by INTI in Argentina. INT in Brazil, in collaboration with the Confederation of Industries, is offering comprehensive technical information services. ISIRI in Iran formulated a constructive policy of not continuing work on in-house investigations beyond essential exploration unless some industrial entrepreneur expressed interest in using the results if the outcome was favorable. The institute in Thailand has assisted companies in several branches of industry to select better raw materials and to improve operating practices. The center for food technology in Campinas, Brazil, organized an internal study course for technical management development and has used interdisciplinary teams on projects. ICAITI in Guatemala has developed good techno-economic services.

Measures to Improve the Performance of Technologic Institutes

Study of the management of institutions in many developing countries, in the light of the successful operations mentioned above, leads to the conclusion that there are several directions for improvement. If they are to live up to their potential as unique sources of technology for local industry, they should strive toward this ideal: an autonomous multi-purpose organization with techno-economic competence, deriving some funds from public sources for non-sustaining activities such as service to small enterprises, but rigorously oriented toward direct service to the industrial community.

Orientation toward industry. Most technologic institutes have a strong bias toward academic recognition. Their boards of directors are often heavily weighted with representatives from universities. Their limited public relations stress technical publications rather than practical contributions to industrial improvement. Their staffs, particularly senior members, tend to be selected on the basis of scientific attainments rather than capability to render services to industry. In extreme cases they become postgraduate institutions of applied research in which the product is published reports instead of improvement of industrial technology.

To effect the needed change in posture, basic alterations are required in institutional philosophy. New goals should be set, and new systems of awards for performance should be worked out.

One naïve assumption about client interest must be corrected. When a new contract research organization is set up, in either a developing or an advanced country, industrial firms do not immediately rush in with problems to solve. Instead, a clientele has to be developed the hard way by systematic and persistent efforts to determine the types of services that will appeal, to submit convincing proof of ability to provide these services, and by demonstrating a capability to handle problems in a satisfactory manner. The idea of having to sell professional services, like any other commodity, is a severe jolt to many neophytes.

Technical management. Most scientifically trained individuals are ignorant of the principles of management. They are unaware of the fact that a technologic institute must be run as a specialized business enterprise. They do not comprehend that all the activities have to be planned, organized, staffed, operated, controlled, and evaluated to ensure the well-being of the organization. Weaknesses in project definition, justification, and control are common.

There is a growing volume of literature on the management of technical programs. While formerly most of the publications were addressed to the problems of advanced economies, in recent years there have been many concerned with those of developing countries. The United Nations Industrial Development Organization has sponsored many of these.

As a first step, there have been many management training courses for senior research executives organized by international organizations, usually under the guidance of foreign experts. The first, a 3-week resident course conducted by an American organization, was held in Cairo in 1964. A number of international organizations are interested in sponsoring such activities. These training programs are most effective if they combine formal presentations by experts with small working groups to study the application of the principles to selected problems. Mention should be made again also of the fact that the food technology center in Brazil organized its own internal training in management.

A related activity is the formation of discussion groups of technical executives from different institutions or firms to consider their common managerial problems, a concept first adopted in the United States in 1923. Technologic institutes can well play a leading role in organizing such groups, from which their managers will derive not only the direct benefits of broadening their backgrounds, but also of expanding their contacts with industry.

Development of an industrial clientele. If a contract research organization is to be successful in its objective to improve industrial technology, it must carry out a realistic appraisal of the services in which the interest of enterprise managers can be aroused.

In developing countries the fact must be recognized that large innovative undertakings will normally be based on imported technology. The reasons are obvious: (a) reliability of the know-how has been proved by its commercial use abroad; (b) the operation can be started much more rapidly than would be the case of local technology which has to be painfully demonstrated to be practical by pilot trials on an increasing scale; (c) the greater confidence of investors, either private or public, because economic and marketing feasibility have already been established in a foreign country. These disadvantages will lessen gradually as indigenous technical skills are improved and recognized by industry.

At the outset, therefore, a technologic institute must resign itself to playing at best a minor role in large new undertakings. There are ways, however, in which it can participate in these events by serving as advisors in appraising competitive sources of foreign technology and particularly in adapting it to local conditions through knowledge of raw materials, suppliers of components, and industrial practices. To serve in these capacities it must develop appropriate skills. It must also promote its relations with industrial entrepreneurs, sources of development financing, and foreign contractors.

If these glamorous projects will seldom be available to them as prime technical sources, what remains as their potential market for services? They must systematically canvas the opportunities for which the cost of foreign expertise cannot compete with locally available skills. The more routine activities such as technical information, analysis and testing, trouble-shooting, and product and process improvement and development should offer attractive possibilities for developing expertise and creating client confidence. If they do not demonstrate their ability to handle problems of these types they cannot expect to have industrial managers refer major projects to them.

To promote this orientation toward practicality, all phases of institutional management must be carefully adjusted. In public relations, which most technologic institutes sadly neglect, the activities should be devoted to those which impress industrial executives. In-house projects should be limited to those in which entrepreneurs can be interested or those that develop new staff skills that are attractive to industry. The reporting procedures should be revised to produce documents in the form that industrialists find most useful rather than as academic publications. All managerial functions must be carried out in a business-like manner.

Technical information services. There is a vital need for improved technical information to assist industry in applying available knowledge to its problems. Emphasis should be on practical utility and not on exotic scientific items. A very promising start has been made in Brazil through the national industrial association which publishes monthly a digest covering managerial and technical information, the latter being prepared by INT, which also issues specialized summaries to several sectors of industry. Information services of this type require subsidy.

Techno-economic appraisals. An essential set of skills which most technologic institutes neglect is the ability to make realistic evaluation of production economics, including operating costs and plant investment, and market potential. These are separate disciplines in which it cannot be expected that technical project heads will have the necessary expertise and objectivity. Proficiency in process engineering and market research are developed largely through the route of experience to supplement training in the pertinent disciplines. Competent market research is vitally important to prevent wastage of resources on unsuccessful ventures; it is most dependable when conducted with a strong technical in-put instead of reliance on economic factors only.

These skills are much in demand by industrial clients to secure unbiased opinion on the feasibility of proposed undertakings. They are used by financial institutions to obtain pre-investment verification of loan applications. Techno-economic considerations should be included to appropriate degree in all recommendations to clients and should be used to evaluate all internal projects to ensure that their objectives are practical.

As examples of the use of these skills, IMIT in Mexico and IIT in Colombia have about 20% each of their professional personnel engaged in process engineering and market research. These are among the most rapidly growing of their services. Also, inclusion of techno-economic specialists in all project work promotes the use of interdisciplinary teams, the most modern form of R & D management.

Techno-economic expertise is particularly valuable for regional and national surveys of opportunities for industrial development. These skills permit a diagnosis of those sectors having the greatest potential for growth and of the incentives and services needed to promote their increased contribution to the economy.

Evaluation of in-house projects. One of the greatest weaknesses of technologic institutes is failure to curb the waste of technical manpower, their primary asset, on unproductive in-house investigations. The projects are commonly selected because of the professional interest of the senior investigator and are often allowed to proceed for years without critical examination of the possibility that the results will ever find commercial application. IMIT in Mexico uses its techno-economic staff as "devil's advocates" to appraise industrial potential after enough exploratory work has been done to define the concept.

Technical and managerial help for small firms. In most countries enterprises of small and medium size account for a large share of national production and employment. In advanced countries competitive forces oblige them to make use of the complex of sources of technical and managerial help to improve their operations if they are to survive. Even this is not felt to be enough, and Canada, Japan, and the Netherlands, for example, are making special efforts under public support to assist them. The most effective technique appears to be the use of industrial liaison officers who, through visits to individual firms, help to identify problems, to suggest practical solutions where they can, but particularly to refer these problems back to appropriate specialists in their organizations. Industrial engineering, a discipline neglected in many developing countries, provides a good background for these functions.

In developing countries the situation deserves strong efforts, supported by the government. The small companies, usually in the more traditional labor-intensive industries, do not have access to the variety of help found in advanced economies. They frequently are not able to identify the underlying problems which are amenable to technical advice, and, even when they can, they do not know where to turn to obtain the information and help.

Services of this kind have been instituted at INTI in Argentina on a very sound basis, coupled with opportunities to improve practical managerial skills through short training courses. These programs cannot be self-supporting because small firms do not feel able to pay the costs, even in advanced countries. They are worthy of strong backing by government to make sure that this vital segment of the industrial complex can become healthier through improved technology and management.

Adaptive technology. There is today much thought being given to appropriate technology for developing countries to increase employment opportunities by reducing the degree of mechanization of processes developed abroad. Technologic institutes can exert important influence by systematic study of opportunities to increase the labor content of such operations, based on comprehensive surveys of current experience in different sectors of local industry.

Recommendations to National and International Development Agencies

In the firm belief that technologic institutes are a prime resource for stimulating industrial development, I offer several recommendations to provide a climate to promote their productivity. These apply both to the governments of developing countries and also to the programs of international agencies which cooperate with them.

It is customary to rely on economists and public administration specialists to develop plans and policies for industrial development. I urge the point of view that these functions become more meaningful if the approach is interdisciplinary, in that the opinions of leaders in technology, industrial management, and finance are incorporated.

1. *A policy on technology* should be formulated to provide public endorsement of its importance to the national welfare. It is common practice to aim for a science policy, or a science and technology policy, but in my opinion these do not give adequate weight to the practical aspects of infusing specialized know-how into the economy. As background information, to give a basis for the provision of appropriate skills required, the government should provide guidance on the level and sectoral directions of industrial growth which it wishes to stimulate.

2. *An assessment of technical resources* should be carried out to provide information on the supply and utilization of technical manpower, the organizations through which it operates, and the facilities available to it. Only through realistic comparison of present resources with future needs can strengths and weaknesses be revealed. Surveys of professional and economic status of specialized personnel have the benefit of revealing inequities in remuneration between industry and public or quasi-public institutions, which should be corrected if the latter are to upgrade quality of staff.

3. *Techno-economic evaluation* of development proposals should be conducted in all cases for which government support is involved, including appraisal of the technology required.

4. *Technical back-up* should be provided for all major public enterprises and for those in which government funding is supplied, because even turn-key plants require continuing improvement in product and process to sustain their operations. This objective can be implemented by stipulating earmarked support of technical activities, both in-house and in local technologic organizations.

5. *Improvement of commodity standards* should be stimulated by tightening federal specifications and by encouraging the standardization efforts of professional and trade groups in the private sector.

6. *Support for technologic institutes* should be increased to encourage them to enhance their skills and spheres of operation, particularly for improving the technology and its management in small enterprises.

7. *Financial incentives to industry* to conduct technical programs should be provided through tax relaxation or cost-sharing to promote technologic improvement, possibly with preferential treatment when local technologic resources are used. Opportunities for stimulating innovation should be encouraged through favorable regulations for obtaining foreign know-how and through patent protection.

8. *Public relations incentives* for new industrial developments should be instituted through honorary awards to enterprises and technologists responsible for these advances.

9. *Industrial statistics* collected by governments should emphasize data that have direct utility in techno-economic development planning, such as the average investment required to create a new job opportunity in different sectors of industry and their respective capital–labor ratios.

10. *Marketing data* compiled by government should be increased to assist enterprises to expand their sales, particularly in regard to product specifications for competitive export markets.

Summary

Technologic institutes are a unique resource for technical help to stimulate improvement in industrial operations. Only in very rare cases do they now approach their potential impact in developing countries, although some of them show encouraging trends. The chief directions for bettering their performance are:

changing their orientation to practical problems rather than academic research; improving managerial practices; actively cultivating an industrial clientele; development of techno-economic skills; providing technical information; stringent control of in-house projects; instituting programs for technical help to small firms; and taking leadership in promoting adaptive technology. They deserve much greater public support for those activities in which they demonstrate their ability to augment their services to industry. Government measures which will encourage their progress are a national technical policy coupled with specific plans for industrial growth; comprehensive assessment of internal technical resources; pre-investment evaluation of development proposals; provision of technical back-up for new enterprises; improvement of commodity standards; increased support of selected technologic programs; financial and public relations incentives for industrial innovation; improved industrial statistics; and increased marketing information, particularly for export trade.

The International Foundation for Science (IFS)— An Institutional Innovation Now in the Process of Formation*

W. MURRAY TODD†

THE idea of forming an international foundation to help finance the research of young scientists of developing countries was first aired at the 1965 Pugwash Conference at Venice by Roger Revelle. The proposal was reiterated and elaborated at the 1969 Sochi meeting. The internationally renowned scientists assembled there told many stories of trials and tribulations of young people of promise who had returned to their homelands from study and research abroad only to discover that the science they were ready and eager to pursue was a will-o'-the-wisp. What sort of international mechanism could be designed to reduce the "brain-drain" from less to more-developed countries, and to facilitate the flow of hard currency to third world scientists to be converted into desperately needed equipment, supplies and literature? Was there a device that would permit a research worker in a less-developed country to concentrate on his research and eliminate the need to hold two, three or even more jobs to make ends meet? How do you convince the bureaucracy of a developing country that science is needed even if it brings no immediate benefit to the country in terms of feeding, sheltering, clothing, or educating its desperately poor people? How do a few scientists, operating in an environment without a scientific tradition, sustain themselves intellectually, and make an impact on the educational, social and economic development of their country?

These are some of the questions that plagued the scholars of both developed and developing country institutions.

In the winter of 1974 the IFS had achieved a degree of acceptance and respectability that was transmuted into modest financial backing by the Swedish and Canadian Governments. The IFS cannot be regarded as a panacea; it will not solve the most intractable problems of development; but if the current exploratory activities bear fruit, the Foundation may result in:

* Originally known as the International Science Foundation, the name was changed to International Foundation for Science in 1971. The text explains why.

† Executive Secretary, Office of the Foreign Secretary, National Academy of Sciences, Washington, D.C.

support of a relatively few carefully selected people, whose research, of international stature, and of relevance to their country's development, would not have been funded from local budgets;

establishment of unique and mutually reinforcing linkages between the scientists of developed and developing countries, which will stimulate international co-operative efforts in research of special use to the developing countries;

improvement of the climate for the support of research by the developing countries themselves;

a new social role for academies, research councils and similar institutions;

a new mechanism for supporting research, slightly less ponderous than are the current intergovernmental agencies which seem so slow and self-constrained.

These are the goals of the IFS. What is its history to date and what are its chances?

Briefly, historical progression begins with the two above noted Pugwash meetings in 1965 and 1969, and then skips to the March 1970 meeting of the U.N. Advisory Committee on the Application of Science and Technology to Development (ACAST), at which the subject was introduced in a joint paper by Drs. Revelle, Marshak, Auger and Salam. Professor Sven Brohult, then Managing Director of the Swedish Royal Academy of Engineering Sciences, happened to be present at this meeting and perceived the idea as one in which Sweden might have great interest.

Professor Brohult took the idea back to Sweden and soon letters went out to a large number of academies, private individuals, foundations, and interested international agencies, inviting them to meet in Stockholm in July 1970. Sixteen representatives of national academies of sciences and similar institutions met at the Hotel Foresta. They came from Western Europe, Africa, South America, South Asia, the Far East and North America. This convocation of scholars and bureaucrats brought forth the following press release, which sums up the results of the meeting:

> The meeting of 16 national academies of science and other scientific organizations and individual scientists from developed and developing countries supports the idea of creating an International Science Foundation (ISF) . . . to help in building up active scientific communities in developing countries by supporting the work of talented individual scientists, or groups of scientists, in research work relevant to the needs of their respective countries.
>
> The special role of the ISF originates in the function of science as a socially responsible, unique intellectual activity of man, which needs for its completeness contributions from all human groups and societies, so that science can contribute its share to the improvement of the human condition.
>
> The foundation will solicit and raise funds from governments, international bodies, commercial and industrial enterprises, and other foundations and individuals for the operation of the foundation. From these funds the foundation will make grants to individual scientists in developing countries on recommendation of their peers in their respective fields.
>
> The ISF grants would be made on an international basis. The foundation, availing itself of the idealism of the world scientific community, would help in this and other ways with the basic objective of supporting pure and applied science in developing countries.
>
> The foundation will be managed by representatives of national science academies and other member organizations. It is expected that the first meeting of an Interim Board will be held before September 1971 in order to formally establish the ISF. A Continuing Committee, consisting of

scientists from developing and developed countries, has been set up to take the necessary steps to establish the ISF at the projected meeting of the Interim Board. It is understood that the ISF—though established as an independent organization—will in its operations be recognized as affiliated to UNESCO.

At the Stockholm meeting it was agreed that there be formed a Continuing Committee of some fourteen people who would carry forward the work of forming the Foundation. The actual task would be undertaken by a Working Group, composed of Pierre Auger, Chairman, Abdus Salam, Roger Revelle, and Sven Brohult. Dr. Adriano Buzatti-Traverso, Assistant Director General for Science of UNESCO, was present at the meeting and offered to provide secretarial assistance and a working home for Professor Auger while the organizing work took place.

If the reader has been reading this account with proper scepticism, he is prepared for the revelation that things did not go as easily as the above lines suggest. To deal with scientists and administrators in conjunction with an intergovernmental organization is not so simple. By October of 1970, when the IFS concept was taken before the General Conference of UNESCO, it was clear that the issue was joined on the question of how independent the IFS should be. The Stockholm group had clearly opted for an independent and autonomous institution associated with UNESCO, but left the form of that association vague. The use of the UNESCO secretariat services, the keen interest of Dr. Buzatti-Traverso, and Professor Auger's long-time, close association with UNESCO, contributed to a situation in which Professor Auger thought it advisable to submit his resignation as Chairman of the Continuing Committee, and all members of the Committee were polled on the desirable type of association with UNESCO. The Committee voted, by a wide majority, in favour of IFS being completely autonomous institution, hoping that it might at some time gain what is known as "Class A Status" as a nongovernmental institution associated with UNESCO.

For the record, it may be useful to note, in very condensed form, the arguments with respect to UNESCO affiliation presented over the year it took to resolve this matter. The advocates of UNESCO affiliation argued that UNESCO could set up the IFS much as it did the International Institute for Educational Planning. This would provide UNESCO privileges for the IFS staff (such as international civil servant status); it would permit duty-free import of equipment into developing nations for the scientists who used the facilities of the foundation to order their equipment; national governments could provide funds to the IFS via the "fund in trust" mechanism which is separate from the UNESCO budget; the developing nations can be presumed to have greater faith in UNESCO than in a new institution funded by the rich countries; and probably the most telling argument—the East European nations, which had signified a reluctance to see an entirely new international nongovernmental institution established, were thought more willing to consider membership and contribute funds to a UNESCO-affiliated organization.

The arguments of those who wanted an autonomous institution tended to minimize the staff benefits and duty-free import privileges as relatively insignificant in comparison to constraints inherent in UNESCO affiliation, such as having nominations for the director of the foundation subject to approval of the Director General of UNESCO, the statutes and working rules of the foundation requiring the approval of the Executive Board of UNESCO, the staff hiring and firing practices subject to UNESCO personnel rules; and complications arising from more than 100 nation members of UNESCO having an influence on the affairs of the Foundation.

Probably the most telling argument was that intergovernmental organizations are not the most flexible and effective patrons of scientific research and that to achieve the intellectual integrity and financial flexibility considered desirable by the scholarly community, the IFS must be independent. The model upon which the proposed IFS organization is frankly patterned by proponents of independent status are the post-World War II operations of The Rockefeller Foundation, which, using the skills and acumen of a few very capable field representatives, managed to sponsor the research of many of the subsequent European Nobel prizewinners of the last two decades. The Rockefeller Foundation's "search and support" activity left a lasting impression on a whole generation of European scientists.

Finally, it did not escape the notice of most of the members of the Continuing Committee that individual Swedes had taken a sustained and profound interest in the concept of the IFS, and that there were reasons to think that Sweden would both appreciate and support an international nongovernmental institution, devoted to development and located on Swedish soil.

Clearly, these arguments are not in direct conflict; but one group opted for the advantages of flexibility and scholarly management, the other, for those of established reputation, known bureaucratic capacity, and presumed greater appeal to governments of the anticipated recipients (the developing countries) and a special group of potential supporters (the East European countries).

Two thoughts might be left with the reader pondering this dilemma. In the first place, should this autonomous institution fail, there is nothing to stop UNESCO (or for that matter, any other intergovernmental organization) from picking up the idea and making it work. Secondly, if the autonomous institution proves practicable, it will contribute something new and unique to international scholarly life; it will be a new institutional invention. It may be noted that this author is not an unbiased observer of this controversy, but feels rather strongly that an independent institution is the right way to try; at least at first.

This controversy occupied the time and energies of the Working Group and the secretariat for most of 1971. None the less, several significant steps forward were taken during this period. The name of the Foundation was changed from that of International Science Foundation to International Foundation for Science. It had become apparent that the first name, rather casually adopted, had a previous owner and there was no desire on the part of either of the two organizations to become confused with each other. Of far greater import, the scope of the scholarship to be supported was expanded to include the full range of behavioral and applied sciences. In January 1972 Professor Brohult, who succeeded Professor Auger as the Chairman of the Working Group and the Continuing Committee, came to grips with the basic problems. Professor Brohult was encouraged by the reception the idea of the IFS had received at a meeting held in fall 1970 in Italy, jointly sponsored by the Rockefeller and the Nobel Foundations, and devoted to discussion of new institutions for scientific and technical development. As a consequence, he had taken soundings among Swedish private philanthropic organizations, and received enough encouragement to urge that the Foundation be created as soon as eight academies in six countries agreed that the endless circle—how to start a granting agency without first having a pool of grant funds available—must be broken. The Working Group, now augmented by Robert Marshak of the U.S., and Friedrich Schneider of the Max Planck Society in Germany, agreed with Professor Brohult that a pilot organization should be formed by erecting a structural skeleton, giving donors a point of contact, and demonstrating clearly the intent of the organizers to see the thing through.

Letters of invitation have been sent to academies and research councils or similar institutions around the world asking them to subscribe in principle to a draft charter, pay an annual membership subscription of $100, and to send a member to meetings of the Board to guide the Foundation. They were also asked to support the idea of the Foundation in their home countries, before their local scholarly and research institutions in the case of developing countries, and before their governmental funding institutions in the case of developed countries.

Over the next year the organizers clarified three critical issues:

The number and types of young scholars in developing countries eligible to make application for grant funds; there are more than any one foundation could help.

The mechanisms and procedures that will provide for each applicant's project to be evaluated in his home country from the point of view of its relevance to development, and abroad by an international panel of peers, from the point of view of its scholarly acceptability; through the use of academies and research councils.

The likelihood of national and international government funding agencies providing IFS with grant funds and operating expenses; and this is still a bit of pending business.

It is evident that many other internal matters also have to be dealt with; for example, the setting of grant criteria, the development of performance criteria by which the Foundation can judge itself and its work, the selection of the first director and his immediate staff, and the selection of a permanent home for the Foundation—although there is a presumption that it will be in Sweden.

The International Foundation for Science should be regarded as an experiment on several planes of action. First, and unquestionably foremost, it is an attempt by scientists of developed countries to cope financially with some of the serious problems besetting their colleagues in developing nations. It is neither a patronizing gesture nor a device to divert the talented scientists of developing nations to arcane and esoteric intellectual pursuits with no bearing on the needs of the country and its people. On the contrary, the Foundation, as conceived, will lend whatever weight it may possess to the enhancement of the highest-quality research that is relevant to the needs of the developing world.

Second, the Foundation is one of several international scholarly endeavours in the process of being launched by academies and research councils in an effort to bring the resources of the international scientific and technological community to bear on problems of the developing world. This is a new role for these institutions, on which they are embarking with some trepidation and great care. But if it succeeds, a new and innovative mechanism of social action may have been discovered.

Finally, the Foundation is perceived to fill a gap left by the activities of private foundations on the one hand, and large intergovernmental agencies on the other. Keyed as it is to support of individual scientists, it may be able to perform a unique service both for science and for development. In this last regard, should the Foundation network of senior world scientists prove viable, it may also become an important adjunct to the major international donor agencies which are badly in need of unbiased scientific advice and review facilities.

*How Can we Exorcise Malthus' Ghost?**

ROGER REVELLE†

THOMAS ROBERT MALTHUS, the parson-turned economist, eloquently stated the long-held and still widely popular view of what we now call the population problem. He believed that human beings would always increase in numbers up to a level set by the available food supply or by enemies and disease. "Gigantic, inevitable famine" stalked in the rear of misery and vice to limit the number of mankind. The neo-Malthusians of our day talk in the same terms, even though death rates throughout the world are lower than they have ever been, and the proportion of the world's human population that is seriously malnourished is less than at any time since the Paleolithic.

Today's unprecedented rates of population growth are a consequence of improvements in the human condition, yet the belief is widespread that this growth is leading to a catastrophe of unprecedented proportions. Although economic and social development are taking place in nearly all countries, the prophets of doom are convinced this situation is temporary and will soon be reversed under the pressure of mankind's increasing numbers.

If we set our sights on the distant future, it is hard to argue with this apocalyptic point of view, simply because long-range forecasting, in our times of rapid transition, is an impossible exercise. But we can say that during the next 30 years the technical potentialities exist not only to feed all human beings, but greatly to improve the quality of human diets. The natural resources available to present technology are sufficient to allow a vast improvement in the standard of living of all the people who will inhabit the earth 20 to 30 years from now. This is not to say that such an improvement in diets or standard of living will inevitably occur. It will depend on the maintenance and improvement of social and economic institutions, and on the growth of cooperation and interdependence among the peoples of the world.

One thing can be said with some certainty; within the next 10 to 20 years, no change in human fertility patterns can have much effect on the dimensions of the world food problem. As we shall see, a reduction in present rates of population growth is highly desirable from many points of view, but such a reduction cannot be a substitute for intensified efforts to bring about economic and social development. Large capital investments and massive transfers of technology are essential elements of such development, though a reduction in human fertility is also an important

* Presented at the Pugwash Conference on Science and World Affairs, Fontana, Winconsin, 1970.

† Dictor, Harvard Center for Population Studies, Cambridge, Mass.

component, as high fertility and rapid population growth have seriously adverse social and economic effects.

The history of human populations gives us hope that in the future social inventions will be developed, as they were in the past, which will lead to a deliberate limitation of fertility by individual couples. But in the meantime we are faced with the overwhelming fact of population growth and its destructive effects on human welfare in the poor countries. Recently, a special Committee of the National Academy of Sciences has examined these consequences.

The Consequences of Rapid Population Growth

Rates of population growth in many poor countries of Asia, Africa, and Latin America are at least half the rates of economic growth and in some cases almost equal the latter. Chiefly because of high birth rates, the ratios of children to adults are very high when compared to these ratios in the rich countries, and the numbers of children and of young people seeking employment are rapidly increasing.

The classical theory of the economic consequences of rapid population growth in the poor countries focuses on these two factors: (1) the high dependency burden (proportion of non-working consumers to workers) resulting from the high proportion of children, and (2) the heavy investments required by the rapid growth in the labor force. During the next 15 years, the size of the labor force would not be affected by a decline in birth rates, because all the new workers who will be seeking employment are already born. But with continuing high birth rates, the total national income must be divided among an ever larger number of persons, and, hence, per capita incomes will be smaller than would be the case with fewer births, and a lower dependency burden. A sharp decline in birth rates would result in higher per capita incomes. However, the calculated effects over the next 20 years, which is a long time horizon for most politicians, are relatively small, compared to the effects produced in different countries by differences in the rates of saving and investment, or in the productivity of labor and capital.

If, instead of focusing on theory, we look at what is actually happening in the poor countries, we see that rapid population growth is affecting human lives in many serious and often tragic ways. Even the agricultural revolution, which is essential if the poor countries are to feed their growing populations, may add to human misery unless birth rates can be quickly lowered. Problems of unemployment and underemployment, urbanization, child development, education, and ethnic conflict are greatly worsened by rapid population growth.

The new agricultural technology is much better suited to some regions than to others. In India, irrigation development is easy to accomplish in the Gangetic plain of Uttar Pradesh and Bihar; it is difficult and expensive in most of the Deccan plateau which covers central India. National farm prices will almost certainly fall because of greatly expanded production in the Gangetic Plain and other favorable regions where the new technology can be successfully applied. The farmers in the

Deccan may then be unable to sell their crops at prices sufficient to pay for the water, chemical fertilizers, and other inputs needed for high-productivity agriculture. They will be forced back on subsistence farming, but this will be insufficient to feed the growing populations of their own villages, even though the *total* national food production will be enough for everyone. Large numbers of poverty-stricken and unskilled countrymen will be driven out either to cities and towns or to more favored agricultural regions, where most of them will become landless laborers. The challenge to policy-makers, either to develop new agricultural technologies for non-irrigated land or to provide employment and a new way of life for these people, is very great.

An extreme example of the rapidly accelerating problems of unemployment and underemployment in the poor countries is seen in East Pakistan, the isolated eastern half of the world's fifth largest country.

In 1961, when the last census was taken, the population of East Pakistan was about 55 million and there were 19 million workers employed or seeking employment. As is customary in labor statistics this number did not include unpaid housewives. Of the persons counted as workers, 16 million, or a little less than 85% (in contrast to about 5% in the United States) were farmers, and 3 million worked outside of agriculture as shopkeepers, skilled and semi-skilled industrial workers, unskilled laborers, fishermen, transport workers, domestic servants, office workers, and in professional and technical occupations. About 1.8 million or 60% of these non-agricultural workers were employed in the villages and the remainder in the cities and towns. Thus approximately 11% of the rural workers had jobs outside agriculture. On the average each farm worker cultivated 1.16 acres. For comparison the average farm worker in the United States cultivated about 75 acres.

In 1985, if present rates of growth continue, the population will be about 125 million and with the same proportion of workers as in 1961 the labor force will contain 42 million people. To produce sufficient food, it will be necessary to increase the intensity of cultivation by about 25%, that is, an average of two crops will be grown on 75% of the cultivated land, as against about 40% of the land in 1961. Thus with the same land/man ratio per crop in 1985 as in 1961 there would be 20 million farm workers, and the number of workers seeking non-agricultural employment would be 22 million. If cities and towns grow extremely rapidly, urban employment might be about 7 million. This leaves 15 million people who would need employment outside agriculture in the countryside. It is hard to see how more than 4 million could find jobs. The remaining 11 million would be either unemployed or grossly underemployed. The rate of unemployment plus underemployment in the entire labor force would be about 25%. Either urbanization and industrialization must proceed much more rapidly than now seems possible, or many more people must be absorbed in the farm labor force, through a considerable increase in the intensity of hand labor on the farms. For example, if the cultivated area per farm worker were reduced to two-thirds of an acre, the number of agricultural workers would be 30 million and the non-agricultural workers in the countryside only 5 million, which would be a higher proportion of nonfarm to farm workers than in 1961, but probably is feasible. Such a reduction in cultivated area per worker would require the substitution of hand labor even for much of the simple farm equipment now used.

A continuing increase in hand labor on farms is said to have taken place in Java with the rapid growth of its rural population, during the past 100 years. It may well have occurred over the past few decades in East Pakistan, but such "agricultural involution" may be incompatible with the modernization of agriculture, which is essential if the people of East Pakistan are to feed themselves.

In times to come, our century may be thought of as that point in history when the way of life of human beings changed from a primarily rural to a primarily urban one. In 1900 no more than a quarter of the world's population lived in cities and towns; by the year 2000 probably more than 50% will be city dwellers. Just as populations are now growing more rapidly in the poor countries than they ever grew by natural increase in the presently rich ones, so the rate of growth of cities is now faster in the poor countries than it ever was during the period of most rapid urbanization in Europe, North America, and Japan. Both rural and urban populations are growing because of the excess of births over deaths, but the cities and towns are also absorbing large numbers of rural migrants. For example, between 1950 and 1960 in twenty-four countries with per capita incomes of less than $250 per year, cities of over 100,000 inhabitants grew 60% more rapidly than the total population. The migration from the countryside was due at least in part to the diminishing size of farms and increasing difficulties of making a living in rural areas. Stagnant rural economies are unable to absorb their own population growth. Although living conditions for many migrants in the cities are appallingly bad, they are probably better from several points of view than in the villages.

In one carefully studied farming region of the Punjab in northwestern India, the rate of migration to the cities during the 1950s equalled half the rate of natural increase in population. With the increase of rural prosperity resulting from the "green revolution", the net migration has diminished by 50%.

In East Pakistan, room must be found in cities and towns during the next 15 years for more than 10 million persons, twice the present size of the urban population. The difficulties of accommodating these numbers in existing cities would be enormous. New cities and towns and new industries must be created on a very large scale. Planners and policy-makers will have hard choices to make in dividing scarce resources between investments to provide industrial and service jobs and construction of housing, roads, water and sewage systems. Other choices must be made between developing many towns of 10–50,000 inhabitants and larger cities with hundreds of thousands or millions of people.

The forced migration from the countryside causes much hardship. Many of the new urbanites are young men between the ages of 20 and 35 who have left their villages to seek employment, leaving their wives and families behind. Census data for urban areas in East Pakistan show about three males for every two females. In crowded rural areas there are about three women for every two men in the 20- to 35-year-old age group, just the reverse of the sex ratio in the cities.

The primary cause of the high proportion of children in the less-developed countries (usually from 40 to 50% of the entire population aged less than 15 years old) is not their rapid population growth, but their high birth rates, which existed even when these populations were growing very slowly. Even before the post-World War II decline in mortality, these countries, because of their high fertility, had young populations.

If and when fertility rates decline in the future, the populations of the less-developed countries will become older, reaching a maximum average age when the populations become stationary at a low mortality level, and the proportion of children declines to less than 30%. There will then be between two and three adults for every child under 15 years of age, instead of the present nearly one-to-one ratio of children to adults. The average size of families will sharply decrease, and the proportion of families with six or more children under 15 will become small. Today, however, the proportion is probably about 20% of all child-rearing families in the poor countries. Studies of the effects of family size show that in families with many children there is more malnutrition and illness, higher mortality, slower physical growth, and less intellectual development of children than in small families. Family size is not the only cause of these effects, but it is an important element in the interacting network of causes.

In eleven villages of the Indian Punjab during 1955–8, 206 out of 1000 children died during the first year after birth, and 95 died between the ages of 1 and 2, in families where the mother had given birth to seven or more living children. In families of two children, infant mortality was only half as great, and only sixteen out of every thousand children died between the ages of 1 and 2 years. In Candelaria, Colombia, more than 47% of young children who had three or more siblings of preschool age were seriously malnourished. In Thailand, 70% of the children whose next youngest sibling was born within 24 months were malnourished, and only 37% of those in families with a younger sibling.

Large numbers of children in the family diminish not only physical growth but also linguistic skills, intelligence as measured by intelligence tests, and educational performance. These elements are to some extent interrelated. For example, heavier children mature earlier, and early maturers do better in school than late maturers. The child's greater contact with adults or with siblings considerably older than himself in small families is probably even more important, because it helps in learning necessary verbal skills and in developing the ability to think abstractly, which underlies most kinds of human problem-solving.

Data from the British National Survey of Health and Development show that in every social class children from small families perform better in educational tests than those from large families. In the upper manual working class, "only" children and those in two-child families scored about 20% higher than children in families of seven or more. The difference in the lower manual working class between "only" children and children in large families was about 17%. In the upper middle class the difference was somewhat smaller, about 10%. Similar differences were found in the average scores of army recruits on general, verbal, and special mechanical intelligence tests, showing that the effects on children of large and small families persist in adult life.

Many large families existed even in the old days of high mortality. The principal change brought about by the modern decline in mortality and the resulting high rates of population growth has occurred between generations; in a sense there has been an explosion in the number of families from one generation to the next. In rural areas where the supply of arable land is limited, this results either in a fragmentation of farms or in an enforced migration of younger sons and their families to towns and cities. The average size of farms in the Punjab of West Pakistan has decreased by about 50% in one generation. The effects of farm fragmentation can be partially overcome by the formation of agricultural cooperatives among the small farmers, but experience shows that this usually occurs only under the impetus of strong government or outside encouragement.

The number of children enrolled in the primary schools of the less-developed countries increased by 150% during the 15 years from 1950 to 1965, and the percentage of children 6 to 12 years old who were in school rose from less than 40% to more than 60%. These marked increases reflected the value placed on education by the people of all classes and income groups of the developing countries.

With rapid population growth, the numbers of children grow even more rapidly than the total population. The need for educating ever larger numbers increases the difficulty of raising enrollment ratios (the proportion of school age children in school) and improving the quality of education. The high percentage of children in the population reduces the amount that can be spent per child out of any given educational budget. Because each cohort of the population is larger than the last, it is difficult to recruit sufficient numbers of teachers from among the adult population.

In 1965 the United Kingdom used 6% of its Gross National Product for education, while Ghana used 5%. But the school age population (5–19 years) was about 37% of the total population in Ghana and 22% in the United Kingdom. Thus Britain used nearly twice as large a percentage of its GNP per head of the school age population as did Ghana. In absolute terms, the United Kingdom, with a Gross National Product per capita of $1800, spent about $500 per child for education, and Ghana, out of a total GNP per capita of $300, spent about $40 per child.

In spite of rapid expansion of education in the poor countries, the total number of illiterates increased between 1950 and 1965, because the number of children in the primary age group rose more rapidly than the number being educated. Educational planners are aiming at a reversal of this situation in the future by raising enrollment ratios to above 90% during the next two to three decades.

Such an increase in enrollment ratios will be extremely difficult to accomplish in less than 20 to 30 years, even under the most favorable circumstances, and it will be much more difficult if the school age population continues to grow rapidly. Calculations for Pakistan show that unless birth rates are reduced, total enrollment would have to be increased by 517% to raise enrollment ratios from 40% to 95% in 30 years. If fertility goes down by 50%, the increase in enrollment would be 200%. A 51% saving in enrollment would thus be attained at the end of 30 years by the assumed rapid reduction in birth rates. The saving would be only 3% at the end of the first 10 years, but it would be 30% at the end of 20 years.

Total educational costs would also be much smaller if present fertility rates were rapidly lowered. This is basically due to the fact that the proportion of children to adults in the population would diminish. In 1995 the percent of GNP required if the 1970 birth rates continue and enrollment ratios are raised would exceed that for a rapidly declining birth rate by 30%. The amount saved each year if the birth rate goes down would be about 900 million dollars, more than four times the *total* expenditures on education in Pakistan in 1970.

The calculation for Pakistan assumes a growth in GNP of 6% per year, or about 350% by 1995. Even with this very high rate of growth, more than 8% of national income would have to be devoted to education in order to accomplish the planned increase in enrollment ratios, unless there is a marked decline in birth rates. Practically no country today allocates such a high percentage of resources to education. If the economy grows at a slower rate than 6%, the increase in enrollment ratios would probably be impossible to attain without a sharp reduction in fertility.

In countries that do not have a homogeneous population, rapid population growth creates or aggravates political and economic conflicts between racial, cultural, religious, and linguistic groups. Numbers are an important element of political power, especially when an attempt is being made to introduce or to maintain democratic institutions. Ethnic groups usually differ from each other in birth and death rates and therefore their relative numbers change with time. The perception of these changes is usually magnified by conflicting groups, far beyond the reality.

Even in the rich countries, intergroup conflicts, involving real or imagined differences in population growth, often occur. Familiar examples are the rivalry between French Canadians and "Anglo Saxons" in Canada, Walloons and Flemings in Belgium, Yankees and Irish Catholics in Massachusetts, Protestants and Catholics in Northern Ireland, "Whites" and "Non-Whites" in the United States, Slavic and non-Slavic groups in the Soviet Union, English and immigrant Asians in Britain. But partly because of higher rates of population growth together with greater poverty, illiteracy, and political instability, intergroup conflicts are more intense and bitter in the poor countries. The civil war in Nigeria between the Ibo and the Yoruba, Hausa and other tribal groups is the most tragic recent example.

Conflicts between Chinese and Malays in Malaysia and Indonesia, Africans and descendants of Indian immigrants in Kenya, Moslems and Hindus in India and Pakistan, Ghanaians and natives of Upper Volta and Togo in Ghana, Whites and Bantus in South Africa, Jews and Arabs in Palestine, and Arabs and Negroes in the Sudan have also had disastrous consequences.

Inter-group relations may be further exacerbated by the large-scale migration from country to the city, or from one region to another, which is one of the results of rapid population growth in most less-developed countries. One typical aspect of this migration is that people of one ethnic, linguistic, cultural, or religious group move into a region where other groups have previously predominated. Ordinary human hostility to "strangers" and "outsiders" is multiplied by the emotions of racism or religious antagonism.

The problems of ameliorating these conflicts have not been solved, and they represent a most serious threat to the existence of many states. In some cases, far-reaching measures such as mass migration or fragmentation of states into autonomous or semi-autonomous smaller units (as happened, for example, in the division of the former Indian state of Punjab into two states, one Sikh and one Hindu) may be the only feasible policy options. But governments can do much by a more even-handed treatment of different groups, providing not only equal but increased educational and employment opportunities and services for all, and by political and legal devices which protect minorities without jeopardizing the basic interests of the majority. These may require rather drastic but pragmatic departures from normal democratic procedures.

World-wide Efforts to Solve the Population Problem

What can be done about the population problem of our time? Over a billion births would have to be prevented during the next 30 years to bring down the world's population growth rate from the present 2% per year to an annual rate of 1% by the year 2000. The task may well be the most difficult mankind has ever faced, for it involves the most fundamental characteristic of all life—the need to reproduce itself. An unprecedented effort is demanded of governments, intergovernmental agencies, and private organizations, yet success will depend on the private and intimate actions of hundreds of millions of individual couples.

In the past, when death rates were high, only a slight pressure for population control was needed to tip the balance between births and deaths. The twentieth-century countries that have nearly stabilized their populations are well developed both educationally and economically. Neither of these conditions exists today in the countries where high rates of population growth are a serious problem. Most of their people are illiterate and nearly all are desperately poor. Their death rates are actually or potentially lower than were death rates in any country prior to the last few decades.

On the positive side, there is evidence that many women and their husbands in the poor countries desire not to have more children, and are eager to use effective birth-control methods that are within their physical and educational capabilities. Recent declines of birth rates in some poor countries, which are experiencing rapid social and economic development, strongly indicate that married couples will limit the size of their families if they believe it is in their interest to do so and if means for controlling their own fertility are available to them. Sample surveys of both urban and rural people in many less-developed countries show that a high proportion of couples having four or more living children do not want any more. On the average the number of live births exceeds by one or two children the number desired. One reason for this is probably that infant and child mortality are still relatively high. For any individual family there is a large degree of uncertainty about the survival of their children, and the excess children over the number desired can be thought of as a kind of insurance against this uncertainty. Another important reason is undoubtedly the ineffectiveness, difficulty, and hardship of preventing births with the methods presently available.

Until very recently, few nations had explicit population policies. Like the movement of a glacier, population changes were barely perceptible from year to year and yet were inexorable in character, seemingly beyond the range of government policy. Only within the last few years, when the vastly accelerated rates of population growth have become apparent to all, have governments recognized the needs and the possibilities of actions to protect their people from the consequences of their own fertility and to effect reductions in fertility. Population policies are thus a new and untested area for politicians and administrators, who have neither tradition nor public consensus to guide them.

Nevertheless, the governments of developing countries are now adopting population policies at a rate and in a climate of world approval unimaginable even a few years ago. During the 1960s fifteen governments in Asia, nine in Africa, and fifteen in Latin America began to undertake fertility-control programs, or to give support to unofficial programs. Among these were the four largest countries of Asia—China, India, Pakistan, and Indonesia—which together contain over 40% of the world's population. Other states that have initiated governmental family planning programs are the Philippines, Thailand, Turkey, South Korea, Iran, Taiwan, Ceylon, Nepal, Malaysia, Singapore, the United Arab Republic (Egypt), Morocco, Kenya, Ghana, Tunisia, Mauritius, Puerto Rico, Jamaica, and Trinidad and Tobago. Most Latin American countries do not have official government policies, but some give support to unofficial programs. These include Colombia, Venezuela, Chile, Ecuador, the Dominican Republic, El Salvador, Honduras, Nicaragua, Costa Rica, Panama, Barbados, and Cuba. The total population of the countries that have or support family planning is 1860 million, 80% of the population of the less-developed world. Several of the rich countries and many intergovernmental agencies including the World Health Organization, UNESCO, UNICEF, and the World Bank are providing financial help and expert advice for fertility control programs in the poor countries as are the Ford and Rockefeller foundations, the Population Council, the International Planned Parenthood Federation and other foundations and international non-governmental organizations.

Most governmental programs are only 3 or 4 years old, and much learning and experimentation remains to be done. It is not surprising, therefore, that they have reached a relatively small number of couples. It is estimated that about 13 million couples have accepted and are now using contraceptives out of a total of some 340 million eligible couples in the countries that have official programs. This is an enormous achievement, but it is also only the beginning of a solution. Much greater effort, more generous financing, and stronger political support at the highest levels of government are urgently needed.

Before adopting a particular policy to reduce fertility, government leaders need to ask themselves several kinds of questions, including whether the policy is politically acceptable to most people, and how effective it will be. But the most fundamental questions are ethical ones: Will the policy enhance the freedom of human beings as individuals, and advance justice for all human beings as members of society?

Clearly, family-planning programs, designed to give full freedom to individual couples to determine the number of their children and the spacing between births enhance the freedom of human beings and, indeed, give this freedom a new dimension, particularly if wives participate equally with husbands in making the decision. Repeal of laws restricting abortion and free availability of medically safe abortions and sterilizations would further increase the freedom of individual women and of married couples.

But the freedom of husbands and wives to make reproductive decisions must be tempered by concern for the rights and interest of others. The first and most obvious interest to be protected is that of the children already born within the family. The birth of additional children may affect them adversely in a variety of ways. Next come all other members of the society whose economic welfare and social well-being are lessened by rapid population growth, and the younger and subsequent generations, whose opportunities will be diminished by the economic stagnation and loss of amenities caused by this growth. Finally the interests of other nations and societies must be taken into account, because all nations ultimately make demands on the same pool of resources.

Governments have an obligation to protect the interests of all these groups against excessive reproduction on the part of individual parents. One of the most difficult of population questions is to think through and justify governmental policies and procedures to accomplish this end.

There is every reason to push family-planning programs as vigorously as possible throughout the world. But experience to date indicates that such programs alone will not bring about a sufficient reduction in human fertility. Easy availability of inexpensive, effective, and acceptable means of preventing birth is highly desirable, but the essential condition for marked fertility reduction is social and economic development. From 1960 to 1967 seven poor countries achieved declines in their birth rates of 20 to 35%: Taiwan, Hong Kong, Singapore, Mauritius, Barbados, Trinidad and Tobago, and Albania. Seven others showed declines of 12 to 19%: West Malaysia, Ceylon, Reunion, Jamaica, Puerto Rico, Chile, and Costa Rica. Several other countries in Latin America seem to be approaching the threshold of fertility decline: Guiana, Venezuela, Panama, and Mexico. In all these cases there has been a substantial measure of economic development and particularly social change, including notable improvements in education, communication, social infrastructure, and life expectancy, and a higher per capita input of capital and technology from the rich countries. The argument for a higher level and greater continuity of development of assistance from the rich countries to the poor ones is overwhelming.

Because of the young average age of human population throughout the world, especially in the poor countries, the attainment of a level of fertility in which each woman just reproduces herself (an average of two to three children per family depending on the level of infant and child mortality) would not result in a cessation of population growth until several decades had passed. If birth rates throughout the world can be brought down to this level (between 15 and 20 per thousand per year) by the year 2010, the earth's human population could level off at around 10,000 million people by the middle of the twenty-first century. This number may seem frighteningly large, but from the point of view of foreseeable resources of arable land, water, air, minerals, and energy, the life support capacity of the earth is probably sufficient to allow a moderately high material standard of living for this number of people.

If birth rates continue at their present high level for another century and a half, the earth's human population could level off at around 50,000 million people, but the leveling process would be a terrible one. It would result from an increase in death rates to match the birth rates due primarily to widespread malnutrition, and the chances of any future improvement would be minimal. The ghost of Malthus would join the Pale Horseman of famine, pestilence, war and death in a triumphant ride over the earth.

On Technical Assistance of the Second Kind[*]

ROGER REVELLE[†]

IT IS characteristic of our time that its most profound problems are global, involving all men, yet the nations into which men have divided themselves have never been as numerous nor as sharply diverse. A hundred years ago, poverty was the common lot of most human beings; today we find two groups of nations staring at each other across the oceans, one rich and one poor, with a growing gap between them. The technological revolution has continuously improved the levels of living in the rich countries; in the poor ones, its principal effect has been to multiply human misery by causing a rise in population at rates higher than ever before.

The growth of populations and the lag in development are interrelated—each worsens the other. Together, they have created a spectre which haunts the poor countries—whether there will be enough food in the future to feed their peoples.

This spectre will not be exorcised in a year or 10 years. The solution will take at least two generations. It must be approached not as in the past, with short-term, quick pay-out programs, or with gimmicks of various kinds, but with realistic, long-continued joint efforts of the rich and the poor countries on a scale never before attempted. The question then arises, how shall these efforts be organized? How can international bodies and the nations themselves best work together to accomplish the enormous task?

In Asia, which contains most of the world's hungry people, almost all the arable land is already cultivated.[1] In this continent, an increase in food supplies could be obtained in part by reducing losses to pests; but mostly by raising production through higher yields per acre, either by increasing the number of crops per year, or the weight of the harvest from each crop, or both. A destructive fallacy of the post-war era has been the notion that agricultural technology of the developed countries in the temperate zone could easily be adapted for use in the poor countries of the tropics and sub-tropics: The standard approach of technical assistance has been, "We know how, we can show how". The fact is, that we do not know how.

The technology for raising yields in each locality must be created through applied local research and this research must be continuous.

* Paper presented at Pugwash Conference on Science and World Affairs, Ronneby, Sweden, 1967.

† Director, Harvard Center for Population Studies, Cambridge, Mass.

[1] President's Science Advisory Committee, Panel on the World Foods. *The World Food Problem* (Washington: The White House, May 1967), Vol. 2, pp. 429–34.

Consider, for example, the Japanese variety of rice, which gives an increased yield of about 10 pounds for every pound of nitrogen applied as fertilizer. It would seem to be a simple matter to introduce this variety into India, where the local rice variety show poor response to more than a few pounds of nitrogen. But the Japanese variety cannot be grown successfully in India, because the hours of sunlight during the growing season are not as long in India as in Japan. A new variety adapted to the Indian diurnal rhythms and resistant to Indian plant diseases is required. Pathogenic viruses and fungi mutate to produce new strains, which virulently infect previously resistant crops. The plant geneticist is hard pressed to modify his plants to keep ahead of their microbial enemies. In the long run, there may even be a question of whether he or the fungi will be the smarter. Similar problems, demanding new technical knowledge and new development, arise in every aspect of tropical and sub-tropical agriculture. In order to make the quantum jump to market agriculture, many cultural and social problems must be solved, as well as technical ones.

It is true that in some problems of development, Western scientific and technical knowledge can be applied directly or can be easily adapted to conditions in the poor countries. This knowledge can be transferred in training programs and through the services of expert advisers. In agriculture and population control, for example, Western technicians can help with the design of irrigation works, fertilizer plants, and contraceptive factories; surveys of soil and water resources; identification and analysis of country problems; methods of collecting vital statistics and other demographic data; and establishment and evaluation of administrative procedures.

In most questions of tropical and sub-tropical agriculture and human fertility control, however, the Western world has little know-how or technology to transfer.

But its scientists and technicians possess the ability to produce new technology through experimental research. A fundamental conclusion is that there must be a new kind of technical assistance. Instead of "know-how, show-how" this kind of assistance must be based on cooperation between specialists in the developing and the developed countries, working together and learning how to solve the wide range of problems, for which at present, there are no answers.

Research and development assistance has two roles. The first is to discover the knowledge needed to carry on assistance more effectively. Here, the results of research are important. Second, research is a teaching tool, understandable across cultural differences, and neutral with respect to previous beliefs and experience. Here, it is the method of research that is important.[2] Thus, an experimental approach is needed, both to learn, and to learn how to teach. For example, in any particular country there may be several different methods of communicating agricultural knowledge to the farmers: radio and television; printed materials; demonstration farms; training a few farmers in each village at a central "academy"; talking to groups of farmers, etc. A similar range of possibilities exists in disseminating information on family planning. In any given situation, it is not obvious which alternative will be sufficient

[2] Arthur T. Mosher, "Research on rural problems" in *Development of the Emerging Countries* (Washington, Brookings Institute, 1962), pp. 78–79.

for the objective, or will give the greatest benefits in relation to cost. It may be desirable to try all methods in a carefully designed experiment, and then to select and use the best single method or combination.

In the early 1940s, average wheat yields in Mexico were 11 bushels per acre, and the country was importing half of all the wheat it consumed. The Mexican Government created a new agency in its Ministry of Agriculture, the Office of Special Studies, with a Rockefeller Foundation man in charge.

In the beginning, most of Mexico's wheat was grown during the dry season to escape the ravages of stem rust. By application of modern techniques of research in plant breeding, new rust-resistant varieties were developed and continue to be developed; these varieties were designed with short, stiff straws to avoid "lodging" of heavy-headed plants, thereby permitting profitable use of high levels of nitrogen fertilizers, and they were insensitive to differences in diurnal rhythms of sunlight and darkness, thus adding considerably to their geographical adaptability. Soil management, weed control, and irrigation techniques were developed. The research progressed rapidly through the use of two and sometimes three crop generations per year.

By 1964 yields had climbed to 37 bushels per acre, a $3\frac{1}{2}$-fold increase, and 95% of the wheat harvest was grown from improved seeds developed in the Foundation project. Mexico has now a surplus of wheat and is beginning to assist other nations by providing its high-yielding wheat varieties, training young scientists, and lending its own wheat specialists. In 1965 an accelerated wheat-improvement program patterned on Mexico's was initiated in West Pakistan.

India bought 18,000 tons of Mexican wheat seeds in 1966, at the time the largest importation of seed stock in history. These were sown and multiplied on several hundred thousand acres in 1966–7; under good circumstances, up to 40% by weight of the Indian wheat harvest in 1967–8 could come from the Mexican varieties; the increase in the size of the harvest will be limited by the availability of fertilizers and water.

An equally dynamic transformation could occur during the next few years in the production of rice, the staple food of most of the world's people, through the use of new varieties developed by the International Rice Research Institute. This is a truly cooperative enterprise, occupying land donated by the Philippine Government, supported financially by the Ford Foundation, directed technically by The Rockefeller Foundation, with a Board of Trustees from six countries, and a staff drawn mainly from the rice-producing nations of South and Southeast Asia, as well as from some developed countries. With sufficient fertilizers and pest control, and timely water supplies, the best new varieties created by the Institute have yields per crop up to five times those of the native varieties now used. The new wheat and rice varieties may very well become major change agents in international agriculture, to match the role of hybrid corn in the developed countries 30 years ago. They have captured the imagination of political leaders in India, Pakistan, and other developing nations.[3]

[3] A. J. Moseman, "The social and economic consequences of world hunger." Unpublished manuscript.

Technical assistance of the second kind is probably essential to accomplish the needed reduction in rates of population growth in the poor countries. Rapid decline in human fertility levels occurred during the last 20 years in Japan and several European countries, but the conditions in these nations differed markedly from those in the poor nations; literacy rates were high, many couples already were deliberately controlling their fertility, and there were strong motivations to limit family size. We need to find, to learn, how to help people to recognize the changes in living conditions in the less-developed countries that will lower the benefits and increase the costs to individual families of having more than two or three children. At the same time, basic and applied research is needed to develop methods of fertility control that are easier to introduce and to use than the present oral contraceptives and mechanical devices.

A set of guidelines has been distilled for technical assistance of the second kind.[4] Such aid takes time; each project should be continued for as long as need be. It is worth offering only if there is a real local desire to accept and utilize it. It should support projects within the capacity of the nation, in order to avoid frustration and capitalize on success.

The people, the government, and trained personnel must be involved in the project so closely that it will be ongoing when the scientists and technicians from the developed countries depart. These scientists and technicians should be selected for quality, personal as well as scientific, and for willingness to serve on a career basis. Emphasis should be on research, but research results should be put into action as quickly as possible. Quality should always be emphasized. Success is most likely where there is progress on a broad front, not in narrow "bottleneck" sectors.

The developing countries lack the research and educational institutions and the educated manpower that are necessary for the sustained creation and application of agricultural technology. It has been estimated that by 1980 in India alone, the number of college graduates with B.S., M.S., or Ph.D. degrees in agriculture will need to be increased from the present 10,000 to 130,000. In addition, 50,000 engineers, nutritionists, food technologists, and economists will be needed by that date, and 75,000 in other professions. From 300,000 to 600,000 sub-professionals will be required. For the less-developed world as a whole, these numbers should probably be multiplied by two or three.[5]

There are not enough trained persons in the less-developed countries to teach the numbers of scientists, extension workers, and other specialists who will be needed. It will be mandatory to add to their human resources the very large resources of the educational institutions, private foundations, and industries of the developed countries.

Some education, particularly of new faculty members, can be given in the advanced countries, but most should be in the less-developed countries, with their

[4] J. G. Harrar, *The Agricultural Program of the Rockefeller Foundation* (New York: The Rockefeller Foundation, 1956), no pagination.

[5] President's Science Advisory Committee, Panel on the World Food Supply, *op. cit.*, pp. 611–14.

unique conditions and special problems. This will require institution building on a very large scale.

Bilateral versus Multilateral Assistance

Many earnest men have urged that the developed countries channel most, or all, of their assistance to the poor countries through multilateral agencies. In this way, economic aid would be transformed, according to one United States Senator, "from a national charity, and an instrument of cold war competition, to an international responsibility", and an end would be put to "the peculiar and corrosive tyranny which donor and recipient seem to exercise over each other in bilateral relationships". This is, no doubt, an idealistic position, but it rests on a grave misunderstanding of the needs of the less-developed countries, and the nature of the assistance they require.

The amount of capital investment in foreign currency required for agriculture alone is about twice what is now being provided. The total need for the overall economic progress that must take place if the problems of population and food supply are to be solved is likewise about double the present flow of capital to the poor countries, especially if their adverse terms of trade, loan repayments, and "brain-drain" (which can be thought of as a kind of reverse flow of investments) are taken into account. Yet there is plenty of evidence from the World Bank's difficulties in finding "bankable" projects that the developing nations, at their present level of education, technology, and social structure, may be unable to absorb investments at a much higher rate than at present. These weaknesses cannot be overcome simply by transferring existing technology or giving advice based on western experience. New technologies must be discovered, and new experience gained, through research, experimentation, and education, in co-operation with the peoples of the poor countries themselves.

This can be accomplished only if research and teaching institutions in the advanced countries become deeply and fully involved over periods of the order of decades and generations, rather than years. It is not enough to engage individual specialists; groups of wise and willing men are needed who are used to working together, who can call on a wide range of resources, and who can enlist large and growing numbers of colleagues, both in their own country and in the country where they are working. These are the characteristics of universities and government departments, and of the great private foundations in the advanced countries. Members of these organizations have demonstrated repeatedly that they can work effectively and happily with their counterparts in the poor countries, without much trace of a "peculiar and corrosive tyranny".

Most technical assistance, be it bilateral or multilateral, has been conducted under the fallacy that the task was simply to transfer knowledge, and this demanded mainly advice and training. In the United States, for example, it was assumed that universities had staff members who knew what to do, and could do it directly during relatively short assignments in the developing nations, or indirectly by teaching students from those nations. If not, they could hire experts on short appointments.

We now know that the universities need to build up their faculties, and the government agencies their research staffs, for continuing involvements between the campus or the experiment station and the field. For example, the Land Grant Colleges of the United States need additional faculty members with a primary long-term commitment to research and teach for international agricultural development. The practical problems of development must be made academically respectable—an accepted basis for promotion and status in the academic community. The faculty members involved must have a chance to work with their own graduate students in the developing countries, and the students must be able to participate in research programs. One of the fundamental requirements is assured funds on a large scale, and for a long term. Most of these funds will have to be provided by the national governments of the developed countries. To this extent at least, bilateral aid must continue for many years to come.

If the international organizations did not exist, they would have to be invented. But both multinational and binational modes of action are needed. The issue is not to choose between them, rather it is to find ways in which they can be related to each other and can effectively interact. The international agencies should play a central role in the formulation and analysis of the problems of the poor countries because these activities will be aided by international comparisons and the governments of the poor countries should have a major voice in conducting them. Planning, coordination, and assignment of responsibility for assistance should also be international; and much "classical" technical assistance (transfer of existing knowledge) can be provided by international organizations. But most technical assistance of the second kind should be undertaken by educational or other institutions in the developed countries and can be most easily supported bilaterally.

*The Agro-Industrial Complex as an Instrument of International Understanding**

ALVIN M. WEINBERG†

Addendum

During the 5 years since I wrote "The Agro-Industrial Complex as an Instrument of International Understanding," the cost of nuclear power has increased alarmingly. Instead of a nuclear power station costing $100 or even $135 per kilowatt electric, we now estimate that a power station using light-water reactors and constructed by 1980 would cost around $600/kWe. Part of this increase is attributable to general inflation; but perhaps $200/kWe is because of other factors. The estimated cost of large evaporators has also gone up but not as much as has the cost of nuclear reactors. Thus instead of nuclear power from large stations costing 2.4 to 5 mills/kWhe, I must now estimate that, in 1980 dollars, nuclear power will probably cost around 17 mills/kWhe. The cost of distilled water from a very large plant would now be estimated at, say, 50 to 70¢/1000 gallons, instead of the 15¢/1000 gallons I estimated in 1968. This grievous escalation in cost has been only partly balanced by improvements in water husbandry. Where I quoted 150–200 gallons as the estimated daily requirement for growing an individual's minimum diet, this may be reduced perhaps to 75–100 gallons in the future.

Despite these disquieting economic developments, I would argue that in certain parts of the world, particularly the Middle East or the Gangetic Plain, the agro-industrial idea still must be regarded with some seriousness. I say this, basically, because there seems to be no alternative, short of reducing the population, to finding new sources of water. To be sure, the time at which water from the nuclear alternative becomes competitive will be extended, but unless population is limited people will have to obtain water at any cost. The nuclear alternative has the virtue of being almost unlimited. One might hope that eventually it will also become cheaper, but presently this must be regarded as a hope that cannot be proved. Of course, if the whole project were written off in the same way Apollo was written off, then much of the optimistic economics would be restored. That the world will some day join to finance such a venture remains a hope for which technological optimists such as I continue to pray.

Recent developments in the technology of nuclear power, in the growing of new, high-yield wheat and rice, and in the technique of large scale desalination, have opened the possibility of combining these three breakthroughs into instruments for helping to develop countries, and for removing human need as a source of international tension. What is conceived are "nuclear-powered agro-industrial complexes" located in coastal desert areas. These complexes would cluster around very large nuclear reactors and desalting plants. Associated with the complex would be a

* Paper presented at the Pugwash Conference on Science and World Affairs, Ronneby, Sweden, 1967.
† Director, Office of Energy Research and Development, Federal Energy Office, Washington, D.C.

large integrated chemical industry, and, where the soil is available, highly rationalized farms or "food factories". Many suggestions for such complexes have been made in the past, particularly at the Pugwash meeting in Udaipur. The most recent suggestion, which has received considerable attention in the United States, was put forward by former President Eisenhower. He visualizes three such complexes producing power, water, and possible chemicals, located in the Middle East, and providing an economic and agricultural base for the rehabilitation of the Arab refugees. The purpose of this memorandum is to set forth the technical basis for such proposals as the Eisenhower plan, and to speculate on the possibilities of its implementation.

The Technical Basis for the Agro-Industrial Complex

The most important gain has been achieved in nuclear power, especially nuclear power based on very large pressurized or boiling water reactors. In the United States, some 40 million kilowatts of nuclear reactors are now under construction or have been firmly committed. The capital cost of these power stations reached the very low figure of $100 per kilowatt last year, but has since risen. It is now estimated that a 1000-MW power station utilizing a water reactor can be bought in the United States for about $135 per kilowatt. This figure does not include the initial inventory of fuel (which may amount to around 30 million dollars) since this cost is included in the fuel cycle cost. Estimates of fuel cycle costs run around 1.2 to 1.5 mills per kWh; the total cost per kWh of electricity from 1000-MW plants ranges now from a low of 2.4 mills per kWh (Brown's Ferry No. 1 and No. 2 plants of TVA) to perhaps 4 to 4.5 mills per kWh for reactors that are privately financed. Reactors built in underdeveloped, new areas undoubtedly would be more expensive; one would be very safe to estimate their cost at twice the U.S. cost.

Important advances in desalting technology, such as the use of fluted tubes and multi-level construction, suggest that an evaporator, capable of producing 400 million gallons of water would cost as little as 30¢ per daily gallon. I would estimate that such evaporators could be built by 1975. The cost of water from a dual purpose power and water system producing, say, 600 MW electricity and 400 million gallons of water per day, might come to around 15¢ per 1000 gallons at the plant, with the power going for perhaps 3 to 4 mills per kWh. These figures are based on public financing, with a 7% fixed charge on the entire investment.

A third, perhaps crucial, development has been the observation of R. Philip Hammond, that by assuring good (but not the highest) recorded yields of new wheats and rices, the amount of water required to grow 2500 calories of grain can be reduced to the remarkably low figure of 150 gallons per day per person fed. This is to be contrasted with the 2000 gallons per day average requirement. It must be recognized that to achieve this very low water budget, the most rational agricultural practices must be exercised. Water must be applied at precisely the right time; fertilizer and pesticides must be available; general good practice must be observed. Yet it does seem likely that Hammond's estimate is substantially correct—viz. 150

gallons per day per person. Thus, if distilled water is available in a coastal desert at 15¢ per 1000 gallons, the cost of water required to produce the food necessary to feed a man a minimum daily diet amounts to only about 2.5¢. This figure is low enough to make such agriculture interesting to underdeveloped countries.

This coastal desert agriculture based on distilled water would require a very high degree of sophistication. It is for this reason that we would describe such farms as "food factories"; this connotes the very strong technical components required to conduct the agriculture successfully. However, the expertise could well be supplied by relatively few high-level experts, in much the same way as relatively few experts have created viable petroleum refineries in underdeveloped countries with natives carrying out the bulk of the more routine labour.

Finally, there are a number of very new interesting possibilities in the production of heavy chemicals electrically. Perhaps most important is the large-scale electrolysis of water with very high current density electrolytic cells. Technology developed for advanced fuel cells suggests that the current density of large electrolytic cells could be increased by as much as a factor of 10 over present practice. Thus, if power costs can be kept low, hydrogen from electrolysis (and thence ammonia for fertilizer) may well become an alternative source of hydrogen for countries that do not possess sources of naphtha or methane.

The Current Middle East Plan

The proposal put forward by Mr. Eisenhower involves construction of three very large nuclear power and desalting complexes in the Middle East. Two would be built in the Mediterranean, possibly near the Gaza Strip and the northern tip of Israel, a third on the Gulf of Aqaba. The three plants would produce around 1 billion gallons of water per day; this represents more than the total flow of the Jordan River and its tributaries. The amount of by-product electrical power is rather flexible, but probably would range around 2 million kilowatts. Most of this electricity would be used in the chemical industry clustered around the energy source.

It is difficult to estimate the cost of such a complex. Leaving out the chemical industry, but including desalting and electrical generating capacity, I would estimate the cost of the three installations as between one and two billion dollars. The water produced from the complex would be capable of feeding about five million people assuming that 200, rather than 150, gallons is required to feed one person each day.

The financing of such a plan is of course a key question. The Middle East Plan visualizes establishing a stock company that would be subscribed to by individuals and governments. Thus all the capital raised for the venture would be equity capital rather than borrowed capital. This greatly reduces the pressure for an immediate return on investment, and increases the apparent economic feasibility of the proposal.

General Observations

Several general comments might be made on the use of nuclear powered agro-industrial complexes:

The *technology* (though not the economics) is almost surely at hand—that is, this is not a case of using tomorrow's technology to solve today's problems. If we decided to go ahead, the first large-scale plant could be built by about 1975.

With future developments in both desalting and reactor technology, one can expect capital costs and operating costs to fall. I would hope that by about 1980 to 1985 desalted water from the sea could be achieved at less than 10¢ per 1000 gallons, with by-product power going for 2 mills per kWh or less.

The Eisenhower plan envisages employment of refugees for building the original plants and then after the plants are built, employment of refugees as agricultural and eventually as industrial workers. Agricultural "food factories" would require expert supervisors; in general, we visualize the operation to be rather analogous to the famous Gezira cotton-growing scheme in the Sudan.

There is no fundamental reason why the project should not involve many countries—for example, the U.S. might provide the reactors, the U.S.S.R. and U.K. the evaporators. Perhaps the locations of the complexes might dictate which countries supplied which component; for example, some countries might help with construction in Israel and others in the Arab countries.

As for financing, I would visualize such projects as having some of the glamour of the space enterprise, and therefore possibly commanding public money that would not be otherwise available. One suggestion is to describe the agro-industrial complex as the "Apollo of the 70s". However, the financing of the projects should be considered from all possible points of view; no reasonable means of support should be rejected.

The possibility of using agro-industrial complexes in other parts of the world must also be considered. The Gujarat Peninsula in India, Sonora in Mexico, and the coastal deserts of Peru are possible sites for such complexes. Particularly attractive is the upper Ganges Basin in India. Here nuclear power would be used not for desalting, but rather for pumping groundwater during the nine months between monsoons. It has been estimated that 10 million kilowatts of electricity used to pump water from, say 2 million tube wells, and to produce ammonia, could eventually create a vast granary that might actually provide adequate food for India during the next 20 years.

Draft Introductory Statement for the World Plan of Action for the Application of Science and Technology to Development, prepared by the "Sussex Group" (for ECOSOC)†*

Introduction

The World Plan of Action for the Application of Science and Technology to Development starts with the premise that the developing countries must have their own scientific and technological capability. This indigenous capability is necessary not only for increasing production, but, what is more important, for improving the capacity to produce. This argument is developed in Section I of the statement.

There are vast gaps in the developing countries between actual production and potential production through the application of science and technology. The underlying problem arises from the international division of labour in science and technology and the present massive orientation of world scientific effort to the problems and objectives of interest principally to the advanced countries. The consequences of the inherent weakness of scientific institutions in the developing countries are exacerbated because advanced-country-oriented science has negative effects through the "external brain-drain" and through the influence which the centres of scientific interest in the advanced countries have on the orientation of scientific efforts in the developing countries. There is, in other words, an "internal brain-drain" of scientific workers in the developing countries, who tend to work on problems which are irrelevant to their environment. The total result is that the composition of the stock of scientific and technological knowledge in general is becoming proportionately less and less suitable for direct use in the developing countries; in addition, the development of new technologies, particularly directed towards producing synthetic substitutes, has adverse economic consequences for the developing countries. At the same time, the developing countries have specific problems of access to foreign technology.

These are immediate causes of the gap between potentials and actual realizations. In addition, there is the fundamental difficulty that the organization of the economy

* Paper appeared originally in Annex II of U.N. publication, Science and Technology for Development: Proposals for the Second Development Decade.

† Members of the "Sussex group": Mr. Charles Cooper (Secretary), Mr. Christopher Freeman, Mr. Oscar Gish, Mr. Stephen Hill, Mr. Geoffrey Oldham, Mr. Hans Singer (Chairman), of the University of Sussex (United Kingdom); Mr. R. C. Desai, designated by the Office for Science and Technology, United Nations Secretariat.

in the developing countries is often unfavourable to the application of science and technology in production.

I. Science and Technology are Necessary to Development

The advanced countries have found that science and technology are extremely powerful tools which help them achieve their national objectives. As a consequence, large resources amounting in many advanced countries to 2% and even 3% of GNP are currently allocated for research and development. Many times this amount are additionally spent in applying the results of this R & D.

By contrast, modern science has not been pursued to the same extent in the developing countries. This has had far-reaching effects not only on their ability to develop indigenous technology appropriate to their local needs, but also in their ability to absorb foreign technology. It has also meant that they have been unable to capitalize on the social role of science as a modernizing force. The world view of science, which implies that the laws of nature can be understood and that this knowledge can be used, is still alien to a large segment of people in the developing countries.

DEFINITIONS

The spectrum of activities covered by "science and technology" is broad. Not infrequently they mean different things to different people. It is therefore important to define what we mean by science and technology.

We have classified the various levels of scientific and technological activity which require systematic support in the development process into four categories. Any such classification is bound to be arbitrary and do violence to the real complexity of society, but some breakdown is essential for serious analysis and policy formulation.

1. At the heart of "the science and technology system" are those activities which are generally known as "research and experimental development" (R & D or NIOKR in the Soviet system of concepts). This group of activities has been fairly closely defined by UNESCO and OECD on standard international conventions, and has been measured in most OECD countries and some of the developing countries. The definitions stress the original and experimental nature of R & D work. However, some countries have included within this category activities which do not properly belong to R & D narrowly defined, such as information services or geological survey work. We have preferred to keep to the strict definition of R & D, and to classify these other activities as a second category of "scientific and technological services" (STS). Expenditure on STS in the developing countries might often be a multiple of R & D expenditures, perhaps by a factor of two, but there are no systematic statistics.

2. We have defined "scientific and technological services" to include the following activities closely related to R & D:

scientific library and information services;
scientific testing and standards services,
museums, zoological and botanical gardens;
geological, geophysical, meteorological and natural resources, survey work, including mapping;
general-purpose social and economic data collection;
technical and scientific advisory, consultancy and extension services, including patent offices, and related activities.

Whereas R & D has been fairly rigorously defined and measured, this is not the case with "scientific and technological services" (STS). Preliminary definitions have been suggested by UNESCO in its publication entitled *The Measurement of Scientific and Technological Activities*.[1]

3. Scientific and technical education and training and popularization of science and technology through the mass media. The UNESCO scheme includes university services and technology education in STS.

4. Application of science and technology in industrial and agricultural production and services of all kinds, including routine design, engineering, production control and marketing, routine medical services, administration and so forth. The UNESCO scheme includes design and engineering in STS.

Policy for science and technology in the narrowest sense is concerned primarily with decisions relating to research and experimental development, but it is our contention that this decision-making must be closely related to over-all strategy for economic and social development. Policy for science and technology in this broader sense is concerned not merely with generation of new knowledge in the R & D system, but also with the dissemination and application of existing and new knowledge throughout the economy, and with the reciprocal interaction between science, technology and the economy. Policy in this broad sense will be concerned with all four of the categories which we have defined.

INDIGENOUS CAPACITY FOR SCIENCE AND TECHNOLOGY IN THE DEVELOPING COUNTRIES

Science as the means of understanding the natural environment, and technology as the means of controlling and exploiting it, are essential to the efforts to increase production in the developing countries and, above all, to make such increases cumulative and self-sustaining. Every "model" of development implicitly presumes the application of science and technology to production in agriculture, industry and services, in whatever other respects the models may differ. The increase of land and labour yields in the agricultural sector especially depends on higher levels of understanding about the environment and on the technological means to exploit it. Industrialization, by definition, means the wholesale introduction of new technology.

[1] Christopher Freeman, *The Measurement of Scientific and Technological Activities*, Statistical Reports and Studies/Rapports et études statistiques (UNESCO, Paris, 1969).

The same applies to other sectors, such as transport, power, etc., and to social sectors such as health and education. In addition, scientific and technological knowledge in principle provides an escape from the limitations of poor natural resources to much larger increases in productive power than could otherwise be conceived.

This is the most obvious sense in which science and technology are necessary to development. But there is much more to be said. Development means more than a quantitative increase in production. An increase in production would in fact be more a consequence of development than its essence. In other words, development is more than growth. In his proposals for the First United Nations Development Decade, the Secretary-General expressed this by saying that the real problem of development was not to increase production but to increase the capacity to produce. This capacity ultimately is inherent in people. It depends on people with the outlook, knowledge, training and equipment to solve the problems posed by their own environment, and thus control their environment rather than be controlled by it. This capacity has grown in the now richer industrialized countries by a historical process, sometimes natural and slow, sometimes planned and even forced, often aided from abroad, but always also indigenous.

The result of this historical process has been a qualitative social change. People in these societies are aware that the environment can be understood and exploited; and a large number of them are trained in the systematic use of scientific method in order to do so. It is this awareness and these skills which are necessary to development and which underlie the application of science to production.

From the outset, then, we reject the idea that the existing international division of labour in science is adequate for development. It provides no basis whatsoever for development; among other things, the less-developed countries must have an indigenous scientific capability.

The idea that there is no problem about the existing international distribution of scientific capability is based on the view that the developing countries can rely on technologies from governments and industries in the advanced countries. In this view, all that is needed is to clear the channels for the "transfer of technology". But the transfer of technology, though necessary, is not an alternative to the development of local science. It is difficult for a developing country without a science and technology capacity of its own, and particularly without the trained people involved, to know what useful technology exists elsewhere, to understand it, to select it, to adapt, to absorb, to repair and maintain, and to operate. Only in the short run can these tasks be done from the outside. As so often in development, it is a case of "to him who has will more be given". If a nation has built up its capacity in the field of science and technology, it is also in a much better position to utilize what exists elsewhere. So all the considerations converge on this priority task of implanting into the developing countries themselves the research and technical capacities required.

In addition, of course, the ability to use transferred technology as the basis of further innovations—which is crucial to maintain competitive efficiency at rising income levels—depends entirely on local scientific capability.

Our starting point is, therefore, that there is a fundamental necessity to build up indigenous scientific capability in the developing countries. This capability is necessary if the people of these countries are to be able to define, analyze and solve environmental problems and escape from the narrow commitment to environmental influences that is characteristic of their present situation. We argue that this local capability is necessary in any case, even if a large number of specific problems can be solved by technologies that have already been developed or could be readily developed in advanced countries.

But the resources of the advanced countries are equally essential. The less-developed countries face many immediate and pressing problems which require a scientific and technical "input". It is quite certain that they will not be able to build up their own scientific capabilities rapidly enough to provide this input in the near- or medium-term future—even if there is a high level of commitment to the development of science. The only way to solve these problems in a reasonable length of time is to mobilize scientific resources from the advanced countries.

In fact, there is probably a reasonable international consensus on these points. The problem is that, in spite of this and in spite of some efforts to apply science to the benefit of the developing countries, so little has actually been achieved. We know that the potential for increasing production in the developing countries is very large; and yet, despite the existence of the scientific means for getting these increases, so little has happened. It is this situation which must be diagnosed and understood as a basis for future action.

THERE IS A VAST GAP BETWEEN POTENTIALS AND REALIZATION

We shall examine four main elements which between them have been highly responsible for the limited impact of science and technology in the developing countries. These are

(a) the weakness of scientific institutions in the less developed countries;
(b) the "weight" and orientation of advanced country science and technology and its impact on the developing countries;
(c) the problems of access by the developing countries to world science and technology;
(d) the obstacles to the application of new technologies arising from underdevelopment itself.

There is, however, an additional factor which must be kept in mind throughout the following discussion. This is the very lopsided nature of the present international division of labour in science and technology. We can give some rough quantitative guides to the international distribution of R & D efforts.

Table 1 indicates that 98% of the R & D expenditure outside the socialist countries is made in the developed market economies. It is possible that the developing countries have a somewhat larger share of world STS expenditure, since the proportionate expenditure on non-R & D STS is probably higher in the developing than in the developed countries. It is also possible that the proportion of world R & D manpower in the developing countries is greater than the proportion of expenditure, since R & D expenditures per scientist (including the salaries paid) are typically a good deal lower than in the developed countries. Finally, the R & D expenditures of centrally planned economies, which are not available to us, should be included to complete the global concept. But, we do not believe that these caveats would lead to much modification of the picture given in Table 1.

TABLE 1. DISTRIBUTION OF R & D EXPENDITURES IN THE WORLD (EXCLUDING CENTRALLY PLANNED ECONOMIES)

Country group	United States	Other developed market economies	Developing countries
Percentage of world expenditure	70	28	2

Source: Calculation based on OECD data for the developed market economies; on UNESCO and Pan American Union (PAU) data for developing economies.

The international division of labour in R & D is profoundly influenced by the national political and economic objectives of the advanced countries. Table 2 shows the distribution of R & D expenditures in the developed market economies by major objectives.

TABLE 2. PERCENTAGE DISTRIBUTION OF R & D EXPENDITURE IN OECD COUNTRIES BY MAJOR OBJECTIVES, 1964

Atomic	Space	Defense	Sub-total	Economic	Fundamental and welfare research	Specific problems of developing countries
7	15	29	51	26	22	1

Source: *OECD Data in The Over-all Level and Structure of R and D Efforts in OECD Member Countries*, (Paris, 1967).

Although the developed market economies make certain expenditures on R & D related to specific problems of the developing countries, these expenditures, on available evidence (from the OECD International Statistical Year for R & D), are very small, amounting to less than 1% of gross expenditure on research and development in all cases.

The extreme lopsidedness of world R & D expenditure and of science and technology efforts, as well as its orientation towards certain major objectives of the advanced countries, do indeed have some beneficial "fall-out" effects upon the developing countries. But, in the main, they are responsible for the operation of the following specific factors accounting for the gaps between realization and potentials.

THE WEAKNESS OF SCIENTIFIC INSTITUTIONS IN THE DEVELOPING COUNTRIES

The global analysis of the international distribution of R & D expenditure shows that, in aggregate, the developing countries make very small allocations to those activities.

At the same time, there is a strong suspicion that the minimal expenditure of the developing countries is also less productive than the concentrated R & D activities of the advanced countries. They are less productive from the scientific point of view in the sense that the output of significant results seems to be small in relation to input of resources; they are less productive economically because the scientific work in question is often of little economic or social relevance to the country's own problems and also because the rate at which results are applied is low. Even the presently meagre effort of the developing countries in R & D yields less than optimal benefits to the countries concerned.

Low productivity is partly a consequence of problems in organization of science in the developing countries. University research is frequently "squeezed out" by heavy teaching and consulting loads; applied work in government institutes suffers from lack of finance, red tape and lack of co-ordination between government departments or even within them. Even where there is an apparent concentration of scientific resources, say in agriculture, this hides a reality in which the total research activity consists of a large number of small projects bearing little relation to one another.

The weakness of scientific institutions in the developing countries extends to survey, testing and data-gathering activities. It is also reflected in the general shortage of scientifically and technically qualified people engaged in production activities. Finally, one immediate reason for the limited application of scientific results is generally presumed to be the weakness of the extension and service types of institutions in the developing countries.

These observations suggest that when qualitative factors are taken into account, the effective use of science and technology resources is even more lopsided than the international resource distribution would suggest, and that the industrialized countries have a vast preponderance. But the matter does not rest there. The sheer weight of advanced-country science, as well as its superior quality, has crucial effects in the developing countries.

The Weight and Orientation of Advanced Country Science and Technology and Its Impact on the Developing Countries

There are three main ways in which science and technology in the advanced countries affect the developing countries. We shall examine them under the headings:

(1) the "internal brain-drain" (2) the "external brain-drain", and (3) the composition of the stock of knowledge and its economic consequences.

THE "INTERNAL BRAIN-DRAIN"

The scientific institutions in the developing countries are weak and so particularly are policy-making and planning institutions dealing with R & D and STS. In addition, as we shall state later, there is very little demand or perception of the need for science and technology in the society as a whole. In consequence, the local influences on the orientation of science and technology in the developing countries are weak.

In these circumstances, the weight and orientation of world scientific effort has a preponderant influence on the way science develops and is oriented in the developing countries. Many observers have noted how scientific and technological activities in the developing countries tend to form an "enclave".

Moreover, it is clear that, even in the fundamental sciences, the orientation of science in the advanced countries is strongly influenced by the major national objectives, objectives to which the scientific efforts of the advanced countries are intimately linked; objectives like defense, space exploration, the development of atomic power and so on (see Table 2). By implication, the orientation of science in the less-developed countries is often influenced and determined by objectives which are external to the countries themselves and which have little enough to do with the requirements of development. Sometimes the aid activities of the advanced countries in relation to science in the less-developed countries have reinforced these contradictory tendencies.

The result is a phenomenon which we shall refer to as the "internal brain-drain", whereby a substantial part of the scientific work going on in the developing countries, in addition to being under-financed and poorly organized, is irrelevant to the environment in which it is being done.

THE "EXTERNAL BRAIN-DRAIN"

A more immediately noticeable consequence of the intensive development of scientific and technical activities in the advanced countries is the rapid growth in demand for scientific workers it generates: the "external brain-drain" is no doubt encouraged in considerable measure by this growth in demand. The "external brain-drain" must also be associated, however, with the incapacity of scientific institutions in the developing countries to absorb and use scientific workers.

The large-scale migration of highly qualified personnel from developing to developed countries is of recent origin. However, the volume of that movement (net) may already be approaching 40,000 per year, and as such, is larger than the movement of assistance personnel from developed to developing countries. Under prevailing conditions, this "brain-drain" is likely to increase over the next decade.

The United States Department of Labor has estimated that 380,000 professionals (as well as about 600,000 middle-level workers) would enter that country between 1965 and 1975. A substantial proportion of those people will come from developing countries, and, in addition, further tens of thousands will be emigrating to other developed countries.

The output from the third-level education is increasing in the developing countries at roughly two to three times the rate of aggregate economic growth; in some countries the difference is substantially greater. Unless some way is found of bringing the employment possibilities for trained people into line with their increasing numbers, educated unemployment and/or international migration will continue or increase.

THE ADVERSE COMPOSITION OF THE STOCK OF KNOWLEDGE

The picture that comes out of the analysis so far is of a relatively very small scientific and technical capability in the developing countries, which is undermined by organizational weakness and also by the varied responses of scientific workers in the developing countries to the mass of attraction that advanced-country science constitutes. The environmental conditions in the developing countries receive scant attention. A consequence is that while the stock of world scientific and technical knowledge is certainly increasing at an accelerating rate, its precise composition is such that there are large gaps in scientific and technological knowledge that would be particularly relevant to the developing countries. The work of the Advisory Committee for the Application of Science and Technology to Development has been concentrated to some considerable extent on identifying these gaps. The list of priority areas where the Committee specifies that new knowledge is urgently required is in one sense a demonstration of the important technical problems which have been left neglected and unsolved by the present concentration and orientation of scientific effort to the political and economic objectives of the advanced countries. Perhaps one of the contrasts is found in the relatively vast knowledge we have of technical development in agriculture in temperate as opposed to tropical regions.

In addition, the stock of scientific and technological knowledge is proportionately less and less directly suitable for use by the developing countries. This is particularly true where knowledge in question is about the application of scientific principles. First, the new technology is not "appropriate" for the developing countries in that it emphasizes production methods which are suitable for capital-rich, and unskilled-labor-short countries, i.e. the richer countries of today. The developing countries by contrast are short of capital and skills, but relatively rich in labor. This discrepancy between the resource-mix for which modern technology is increasingly designed, and the actual resource-mix in the developing countries places them at an increasing disadvantage. Secondly, the available technology emphasizes production on a large scale whereas the initial markets of developing countries (even including their realistic export markets) are usually small in economic terms. Thirdly, product design of plant, equipment and consumer goods emphasizes the needs of the richer industrialized countries. Finally, a very great deal of world scientific and technological effort is concentrated in industries which simply do not exist in the developing countries, and which will not exist there for many years to come.

But the problem is not only that the needs of the richer countries are dominant, but that the products of scientific and technological progress which result from this concentration are often such that they exert harmful "backwash" effects on the economies of the developing countries. Apart from the "brain-drain", the development of synthetics replacing the natural raw materials produced in the developing countries is an important case in point.

Probably about $1000 million per annum are being devoted to R & D on synthetic materials (plastics, fibres and rubbers) in the chemical industries of the advanced countries. This is almost equivalent to the entire expenditure on research of all kinds in the developing countries and is, of course, very largely devoted to new materials of primary interest to the advanced economies.

When the advantages and benefits of further development of synthetic substitutes are considered, the harmful effect on the producers and exporters of the natural primary commodities thus displaced is not normally taken into account. The results are all too apparent: the shares of such natural commodities as rubber, cotton, tin, vegetable oils in total world consumption and trade have declined rapidly, partly as the result of research and development devoted to the economy in the use of such materials and/or the development of synthetic substitutes for them. Work on the development and improvement of natural primary commodities of special interest to the developing countries does not receive the emphasis it deserves.

THE PROBLEMS OF ACCESS BY THE DEVELOPING COUNTRIES TO WORLD SCIENCE AND TECHNOLOGY

A further problem is that the developing countries have highly imperfect access to the body of world scientific knowledge and also to world technologies.

Easy access to sources of information, and effective "coupling" with these sources, are essential to the efficient functioning of the science and technology system in any country. This "coupling" function has to be done across national and cultural boundaries, which presents exceptionally acute problems. Thus, studies of information flows in the R & D process generally conclude that information derived from formal literature search and from formal information systems account for relatively small proportions of total information inputs. The informal communication network is of critical importance, including direct personal contact, telephone calls and correspondence. This is one of the critical advantages of the industrialized countries in the workings of their science and technology systems, and it poses special problems for the developing countries, which have to establish these informal links with the scientific community in the advanced countries.

Access to world technology, on the other hand, raises special problems. Much of the technology in question is in private ownership, that is, it is patented or at least secret. In general, the companies owning the technology "release" it to the developing countries only if they are able to make direct investments there. The companies show a marked preference for direct investment as a means of exploiting the technological advantages in the developing countries, rather than entering into agreements with independent firms in the developing countries themselves. The main reason for this seems to lie in the lack of capital and skills in potential counterpart companies and also in the risks of inefficient operation of the new techniques by independent enterprises in the developing countries. The net results are that the flow of proprietary technologies to the developing countries are dependent on their capacity to bring in foreign investment (which is limited) and that the development and competition of domestic industry is hampered by lack of access to new techniques.

Even when technology is non-proprietary, there are problems of access for the developing countries. They have to obtain technology in an embodied form, that is, by importing capital goods and/or building up capital goods industries of their own. In these tasks, they are hampered by lack of domestic savings and lack of foreign exchange. The latter is, in part, the result of world science and technology effort which, as we mentioned before, has tended to reduce the earnings of the developing countries from the exports of primary products. On the other hand, the volume and costs of importing capital goods have been going up, partly reflecting the increasing sophistication of embodied technology. The capital goods which the developing countries import are not only more expensive *vis-à-vis* their exports, but, in most instances, malsuited to their resource endowments.

UNDERDEVELOPMENT AS A BASIC OBSTACLE TO THE APPLICATION OF SCIENCE AND TECHNOLOGY

While these various factors are important in explaining why it has been so difficult to exploit science and technology for the benefit of the developing countries and why the growth of science and technology has had adverse consequences for them, they are really only proximate causes. The real causes lie deeper, in the nature of underdevelopment itself. In brief, many of the structural and organizational characteristics of the developing economies are antithetical to the application of science and technology and, by the same token, prevent the development of what might be termed a "realized demand" for scientific and technical knowledge. It underlies both the limited transfer of technology into local industries in the developing countries and also the weak development of local scientific institutions and their marked susceptibility to orient their activities in line with external influences. This is a particular aspect of the "vicious circle of underdevelopment": the resolution of many of the problems of the developing economies requires the application of science and technology to production, yet the conditions of underdevelopment itself tend to limit the possibilities for their application.

Hence, while science and technology are certainly necessary inputs for development. their application in the developing countries nearly always requires certain important structural and developmental changes as concomitants.

A Major Reorganization of World Scientific Effort is Needed: Objectives, Strategy and Implementation

On the one hand, we have argued that scientific and technological efforts in all four of the categories defined in Section I are necessary to self-sustained development and to creating the capacity to produce. This requires an indigenous science and technology capability in the developing countries and a major reorientation of world scientific effort.

On the other hand, the analysis shows that this will not happen automatically. The application of science and technology in the developing countries cannot be left to chance. If it is, nothing constructive will happen or, more probably, the ensuing development of world science will have negative effects on the developing countries. There are self-reinforcing tendencies in the international scientific system which work against the development and exploitation of science and technology in the developing countries.

In other words, the question must be approached consciously and systematically; scientific and technological effort must be directed to the objective of development in a planned way, just as the application of science to space conquest has been planned.

But the planned application of science and technology to the problems of the developing countries is particularly difficult because the planning has an international dimension; it will not succeed without international co-operation. The necessity for international action arises simply because the developing countries cannot solve the problems arising from the present international division of labor in science and technology by their own efforts.

This has important implications for the discussion of policies to resolve the present crisis. In order to exploit the potentials of science and technology in the developing countries, various policies are needed both in the developing countries themselves and in the advanced countries. By implication, these policies are linked; they are all equally necessary and the effectiveness of any one of them will be vitiated if the others are not also undertaken. The main policies needed fall into four groups:

(a) policies in the developing countries to build up their indigenous scientific and technological capacity in an orderly way;
(b) policies in the advanced countries to support these efforts and particularly to reorient world scientific and technological efforts;
(c) policies to provide the developing countries with better access to world science and technology; and
(d) policies to resolve the problems of economic organization which prevent the rational application of science and technology to production in the developing countries.

DEVELOPING-COUNTRY EFFORTS TO BUILD UP INDIGENOUS CAPABILITY IN SCIENCE AND TECHNOLOGY

Indigenous scientific capability is essential for creating the capacity to produce in the developing countries. The present weakness of their scientific efforts reflects a random growth of scientific activity, which is inevitable if things are left to take their course in societies where there are no pressures of demand for scientific and technological knowledge. This means that the development of science and technology in the developing countries will have to be much more carefully planned than in the past.

The objectives that will have to be met as a basis for this planned buildup may be inferred from the preceding analysis. They are

(a) the formulation of effective science policies;
(b) the provision of highly and appropriately qualified manpower for R & D and relative activities. The university system must be able to provide science education to the postgraduate level and to relate research to the context of development;
(c) the execution of R & D supported by necessary STS;
(d) the application of new techniques in production, and the promotion of new attitudes based on scientific method among management, workers and the total community.

INSTITUTIONS FOR FORMULATING SCIENCE AND TECHNOLOGY POLICIES

The necessity for policy-making institutions is implicit in everything we have said and particularly because the scientific and technological effort has to be planned in the developing countries. Science policy institutions must not be allowed to become merely a façade for scientist pressure groups who, under the guise of national commitment, are really mainly interested in getting more resources put into their own fields of interest.

What is needed is an institution or institutions that can fulfil the following functions:

(a) coordination of science and technology strategies at senior executive levels of government with national economic and social planning;
(b) coordination with the science and technology strategies of other countries. National science and technology policy cannot fail to take into account what other countries are doing, or fail to consider the implications their programmes may have for national programmes;
(c) the establishment of scientific activities which relate to the development effort of their country and the orientation of scientific activities in directions most relevant to national development;
(d) integration of "coupling" between research effort and the process through which this effort is translated into economic applications;
(e) establishment of policies towards technology transfer.

The requirements for science policy institutions have been analysed in some depth by UNESCO; their work is an invaluable guide to the institutions which are required. Unplanned science expenditure in the developing countries will simply increase the "internal brain-drain". Planning should relate science and technology activities to development objectives and should specifically take account of complementarity between indigenous R & D and the importation of technology.

The Reform of the Education System so as to Provide Appropriately Qualified Workers for Scientific and Technological Activities

The accelerated development of science and technology will put pressure on educational institutions and particularly on postgraduate education. There is little doubt that the real constraint to the development of indigenous R & D in the developing countries is the scarcity of manpower. This reflects acute limitations in the university organization, which undermine the possibility of maintaining very high growth rates of R & D expenditure.

The problem in the developing countries is not only that the stock of scientifically qualified manpower (and particularly R & D manpower) is very small (which makes high growth rates easier to conceive), but that the institutional capacity to produce more is also small (which makes the high growth rate very questionable even though the absolute increases are not large). Data on proportions of qualified personnel are scanty; and there is very little information on the past growth of the stock of scientific and technical manpower. We can, however, get some idea of the potentials for increasing stocks of total scientific and technical manpower by looking at proportionate additions to stocks of scientists, engineers and doctors. Some very rough data available for Middle Eastern and North African countries suggest that proportionate additions of 8 to 12% are fairly typical. Data on postgraduate outputs are much more limited than general data on graduates. Fragmentary information from OECD studies suggests that additions of postgraduate trained manpower to stocks of scientists engaged on R & D are about 5 to 10% in certain southern European countries.

In other words, in spite of the small absolute size of the stocks of scientists it seems unlikely that it would be possible to increase these stocks by much more than 10% per annum over 10 years unless there is a reform of the universities or unless educational facilities in the advanced countries can be used more intensively without increasing "brain-drain" losses. This latter possibility requires very considerable changes in the present world scientific content.

University reform leading to an increased rate of output of graduates and postgraduates for R & D is evidently necessary. It is, however, a slow process. It tends to be slower and more arduous in those countries which have reasonably large and relatively long-established university systems, as is the case in Latin America, and India.

THE REFORM OF R & D AND STS INSTITUTIONS

But even if reforms in the education system are carried through, there will remain the problems of the capacity of existing institutions to increase their rate of performance of R & D and STS work. In other words, manpower availability is not the only constraint. In the short run, at least, there are organizational problems whose precise impact it is hard to measure, but whose effect might be considerable:

(a) The rate of execution of R & D in the higher education sector is very limited because of heavy teaching loads, multiple employment of academics, administrative pressures and so on. These difficulties have run so deep that many academic scientists in the developing countries virtually stop research activities and often lose contact with developments in their disciplines. Part of the problem would be solved by making sure that a relatively high proportion of new research manpower goes into the universities. But the deadweight of established professorial staff would remain, who would probably find it very difficult to go back to research work after years away from it.

(b) The government sector accounts for most remaining scientific and technological activities in developing countries. Here there are some serious organizational disorders arising mainly from the fact that scientific and technical workers are under civil service organization, which is often antithetical to the conditions for doing scientific work. In particular, the growth and coordination of R & D and STS activities are complicated by inter- and intraministerial rivalries. Once again reform is a slow process.

Thus institutional reform is necessary not only to increase the supply of scientific workers, but also to create the capacity to "absorb" them and to provide them with conditions for getting scientific and technical knowledge applied. Both types of reform are slow moving in general.

In short, increased R & D and STS expenditure would be folly if there were no reform of the institutions for carrying out these activities. The organizational problem in these institutes and services have to be solved; otherwise, the additional financial resources devoted to them will largely be wasted. It is relatively easy to increase expenditure on science and technology, but there is a grave danger that the increased expenditure will leak away in a kind of "scientific conspicuous consumption".

INSTITUTIONAL ARRANGEMENTS FOR THE APPLICATION OF TECHNOLOGIES

The fundamental problems of getting technologies applied very frequently lie in prevailing forms of economic organization. At the same time, however, successful application depends on an adequate system of institutions, like extension services, concerned with encouraging the application of technology and on adequate coupling between these institutions and all other STS activities. There must be firm connections between every link in the technology application chain. Thus:

(a) technologists, able to understand the contributions of fundamental research, but familiar with industrial needs and problems, should provide liaison between fundamental and applied research functions;
(b) as close a contact as possible should be encouraged between fundamental research scientists, technologists, applied scientists, extension officers and potential industrial and agricultural users;
(c) liaison and extension officers are also needed to help potential industrial and agricultural users to recognize their technical needs and problems, and to relate applied scientific research to those needs and problems.

While all these service activities are absolutely necessary, their tasks will be exceedingly difficult in social circumstances where various systems of authority and tradition guide individual behaviour and decisions. The business of getting technology applied is much easier where people are able to consider problems of production—other problems—in a rational and scientific way. There is, therefore, a strong case for large-scale programmes to bring an awareness of science and the scientific method to the community as a whole, and to encourage the participation of the community in the actual process of development of new modes of production.

QUANTITATIVE EXPENDITURE TARGETS

At present the developing countries spend on the average about 0.2% of their GNP on R & D proper. This is the best available estimate; it is probably on the high side.

We propose that in the aggregate, the developing countries should aim to reach R & D expenditure levels of 0.5% of GNP by the end of the decade. Given the assumption of an aggregate GNP growth rate of 6% per annum, this would imply an average annual increase of just over 15% in real expenditures on R & D over the decade.

We have expressed the target in terms of R & D expenditure, because the statistical data are best for these activities. It is, however, essential to bear in mind that there should be a similar—or possibly even a great—increase in STS in general and in all other activities listed in Section I. Expanding R & D by itself will not be economically useful.

The target probably represents the maximum feasible growth of R & D expenditure when manpower constraints are taken into account. In an initial phase while the present underfinancing of STS is corrected and when new institutes are being built, expenditures may rise more rapidly; but, on the average over the decade, they will be limited by the availability of R & D manpower.

At the same time, in setting this target, we wish to underline the fact that we presuppose that the various reforms of science and technology institutions in the developing countries discussed above will be undertaken. Unless they are, the expansion of expenditure will be impossible. Or, if there is an increase in expenditure, it will merely exacerbate existing problems and be a waste of scarce resources.

The target is an aggregate one for all developing countries; it cannot be applied to any specific country without qualification. Some countries have already surpassed this level of R & D expenditure; others, like India, are closer to it than the average. Obviously the target will have to be disaggregated in order to make sense in operational terms.

An examination of the operational implications of the target will certainly reveal problems of "sub-threshold" effort; that is to say, in small poor countries the feasible expansion of science and technology activations will result in a situation where the concentration of science and technology activities will be very small, probably far too small to be productive. This question has major implications for implementation. First, it points up the importance of a carefully planned orientation of effort in the developing countries. Secondly, it suggests that regional co-operation between some groups of developing countries (particularly in Africa) is a *sine qua non* for productive scientific and technological effort. In spite of the acute problems to which regional co-operation gives rise, it is essential in many cases.

Clearly, in order to attain an aggregative target of this magnitude, the developing countries will have to initiate considerable changes of a generally difficult kind. But even the major effort that this implies will only have a limited effect on the international division of labour in science. In terms of R & D expenditure, the target we have set would mean that at the end of a decade, the developing countries might account for some 4 or 5% of world expenditures outside the centrally planned economies.

In other words, a great many of the problems associated with the present concentration of world scientific effort would remain. There would still be forces making for "internal" and "external" "brain-drains"; and there would still be a preponderant scientific effort directed to economic objectives in the advanced countries which have negative or "backwash" effects on the developing countries. In all likelihood, the various effects of the international division of labour would in fact vitiate much of the effort the developing countries might make themselves, and could simply lead to a waste of valuable resources in those countries.

Thus, while increased domestic effort of the developing countries is essential if scientific and technological potentials are to be exploited there, a major effort by the advanced countries is no less necessary.

The Reorientation of Scientific and Technological Activities in the Advanced Countries

The individual efforts of the developing countries will come to very little unless there is a parallel effort in the advanced countries, both to provide direct support to the buildup of science in the developing countries and to reallocate their own scientific resources towards new objectives related to the problems of the developing countries. Advanced country assistance may take multilateral or bilateral forms.

Direct Assistance to the Buildup of Indigenous Scientific and Technological Capability in the Developing Countries

A part of the advanced-country contribution to the international effort will be the provision of direct assistance. This should take the form of transfer of physical resources to meet the science and technology requirements of the developing countries, and technical assistance programmes, namely sending experts abroad to advise on science and technology buildup in developing countries and to do R & D work there; and training scientists from developing countries in advanced countries.

The physical resources transferred to the developing countries must meet the following objectives:

(a) They must help to overcome domestic financial constraints which, quite apart from manpower and organizational bottlenecks, might jeopardize the attainment of developing countries' expenditure targets. In other words, the calls on investible resources might be so considerable that resources available for investment in science and technology fall short of planned investments and a "science and technology savings–investment gap" might eventuate. Thus, resources transfers might be necessary to meet local requirements, for example, on setting up new institutes.

(b) They must help to overcome balance-of-payments constraints on the import of expensive items of equipment from the advanced countries. Again, the calls on national foreign-exchange resources for other purposes may constrain the possibility of importing equipment. This might itself restrict the possibility of attaining expenditure targets or, more probably, result in a change in orientation of effort from planned orientations.

(c) They should enable the developing countries to contract for R & D and other science and technology services to be performed for them by institutes and companies in the advanced countries. In this way, demands for scientific and technological inputs arising in the developing countries themselves would exert some influence on the orientation of international scientific effort.

The technical assistance programmes which will be required as part of the over-all effort have to include all those activities designed to help the developing countries meet the conditions which are necessary for the accelerated and planned buildup of indigenous science and technology. Technical assistance activities would then cover the following: advisory work and training on university reform; advisory work and training in "science policy" and in science organization; advisory work on the organization of R & D and science and technology institutions in general, and training of research workers and other scientific and technical workers in the educational institutes of the advanced countries.

In our view, generous advanced-country technical assistance to the science and technology activities of the developing countries is likely to be absolutely necessary if the developing countries are to accomplish the organizational reforms and the setting up of new policy-making institutions which are essential for the orderly attainment of the R & D expenditure target.

Technical assistance in the form of training of scientific personnel from the developing countries involves some particular issues, issues that we deem necessary to spell out in more detail.

The training of highly qualified scientists and engineers is at once the most useful and, at the same time, dangerous activity the advanced countries could undertake. It is useful because the manpower constraint is likely to be crucial to the development of science and technology in the developing countries. It is dangerous because it may simply add to the outflow of trained personnel.

This danger is so pervasive that we are forced to set what we regard as an essential precondition. Training programmes will only make sense in a changed world scientific context, where:

(a) the advanced countries themselves are developing major science and technology programmes for the developing countries and, therefore, have redesigned their scientific education systems so that they provide an education that is relevant to the problems of the developing countries;
(b) the concept of "science for development" is fully recognized by the scientific community and properly built into their incentive structures; and
(c) the scientific institutions of the developing countries are themselves reformed and properly motivated so that the newly trained scientists have something worth going back to.

In any other context—for instance, within the present world context—the increased training of scientific manpower in the advanced countries may simply undermine the efforts of the developing countries to build scientific institutions. The technical assistance aspect of advanced country efforts thus also brings us face to face with the more fundamental necessity for a reorientation of international scientific effort.

A Quantitative Target for Direct Assistance to Science and Technology in the Developing Countries

There is no better theoretical basis for calculating this target than for the others. We have, once again, attempted to arrive at a feasible scale of activity by looking at the discernible constraints. On this basis, we have concluded that the advanced countries might be called upon to transfer resources equivalent to 0.05% of their GNP to the developing countries for direct support of science and technology.

Assuming that the general aid target of 1.0% of GNP is reached, this would represent 5% of total aid in the target year. If all official aid amounted to say 75% of the total flow, it would represent some 7% of official aid flow and would absorb about 12% of the increase in official aid between now and the target year. Given the special importance of the other components of official aid, this would seem to us to be the maximum feasible science and technological component in aid, although our judgment here is based on guesswork.

The Reorientation of Advanced Countries' Scientific and Technological Efforts

The reorientation of advanced countries' scientific and technological activities plays two essential roles in the programme as we conceive it. First, this reorientation is necessary in order to overcome the influence of present patterns of effort on such problems as the "internal" and "external" outflow of trained personnel. The buildup of indigenous science in the developing countries will be undermined unless this can be achieved. Secondly, the reorientation is necessary if the composition of the stock of knowledge is to be significantly changed and if some of the negative effects of advanced-country science and technology are to be overcome. Particularly in the light of these "backwash effects", we do not regard the advanced-country efforts to reorient their scientific activities as part of aid programmes properly so called. They should be viewed rather as a compensatory attempt to correct the irrationalities that arise from the present international division of labour in scientific work, and particularly for the negative consequences which scientific and technological advance has had on the developing countries.

The main objectives of the expanded programme must be to fill the gaps in knowledge relevant to the environment in the developing countries. These gaps have resulted from past orientation of world science to the political and economic objectives of the advanced countries.

The priority areas selected by the Advisory Committee show just how serious these gaps are; they also provide the main immediate objectives for the expanded R & D and STS programmes. We presume, therefore, that a large proportion of the advanced-country resources which will be devoted to scientific and technical problems of the developing countries will be concentrated on the areas for "concerted attack" designated by the Advisory Committee and discussed in detail in the World Plan of Action for the Application of Science and Technology to Development.

In addition, the last two decades have witnessed a great emergence of new technologies which promise possibilities of fruitful application to the developing countries, somewhat parallel to the sharp gains in the fields of medicine and public health brought about by the discovery of antibiotics and DDT in the prior years. Advances in space technology have opened up possibilities with positive implications for agriculture, resource surveys and education. Advances in computer technology promise economy in skilled manpower and management resources as well as gains in quality control. Nuclear energy offers to the arid areas of the world desalination and large-scale agro-industrial complexes. Transistors and miniature circuits have made several industries less capital-intensive and, therefore, easier for establishment and successful operation in the developing countries. The exploitation of the resources of the sea-bed is receiving increasing attention. A large volume of research is being devoted to discovering more practical means for the control of human fertility. It is not yet certain that these discoveries will mean a technological "breakthrough" in the development process of the developing countries. Considerable technical and economic research is warranted in both the developing and advanced countries specifically directed to their meaning for, and application to, the problems of the developing countries.

In essence, what is needed is an attempt to change the composition of the stock of knowledge so that it is relevant to the developing countries. In particular, this means a redirection of resources of science and technology towards the products, raw materials, and methods and scales of production of particular importance to the developing countries. There are really two aspects to this problem: one is to change the spectrum of technologies now being developed and made available to developing countries, so as to add to that end of the spectrum where input requirements are closer to the pattern of resource endowments in the developing countries. The second aspect is to enable developing countries to make better choices, selections and adaptations in the existing spectrum, largely again with a view to enabling them to find more appropriate mixes, which will satisfy their desperate need for reducing unemployment and for spreading the participation and benefits of development more widely among their populations.

At the present time, it would not be far wrong to say that everything conspires to prevent the developing countries from selecting, even from the existing spectrum, the capital-saving and employment-generating technologies that may be optimal. The consulting firms which they employ on preparing projects, making feasibility studies, drawing up specifications and contract provisions, examining tenders, etc., are all steeped in the technological outlook and traditions of the richer countries. The same is true also where local officials or local technicians are available and employed.

The advocacy of greater application of labour-intensive technology as an urgent need in developing countries, both in the sense of a better selection from the existing spectrum and the development of a better spectrum, must not be taken to an extreme. The fundamental requirement is economic efficiency in terms of the resource endowment of the developing countries. We believe, however, that the scope for the employment of efficient labour-intensive technologies is considerably greater than at present.

At the same time, some action is required on the specific adverse consequences which the concentration on synthetic substitutes has had for the developing countries. A purely negative approach to this question would do more harm than good. Attempts to prohibit or "police" research on synthetics would almost certainly be both backward-looking and ineffective. The new materials have properties and uses which are of immense benefit for the whole world economy, and potential applications of value to the developing countries. The primary aim of policy should, therefore, be the positive and constructive goal of building up the scientific capacity of the developing countries so that, on the one hand, they are able to strengthen the competitive position of natural materials through advanced research and, on the other hand, are able to develop a new range of applications for synthetics, which are principally of interest to the developing economies. Many of these have in fact the resource endowments which would enable them to become centres of the petro-chemical industry, and of production of the synthetics. A policy of restricting the developing economies exclusively to natural materials would be suicidal in the long run. They must themselves have the scientific and technological capacity to participate in the synthetic revolution, while simultaneously exploiting their natural resource endowment.

A Quantitative Target for the Reallocation of the Scientific and Technical Resources of the Advanced Countries

The reallocation of advanced-country science and technology towards the problems of the developing countries should be treated as an essential means of overcoming the adverse consequences of the present international division of labour in science and technology; for example, the outflow of trained personnel and the development of synthetic substitutes and methods for economizing on the use of primary materials. Attempts have been made to quantify the effects of these factors. We believe that the sum total of what is being proposed here in terms of reallocation of effort will do no more than compensate the less-developed countries for the combined effects of these adverse impacts.

In examining this question against the necessity to get a substantial change in the orientation of world science and technology, and to overcome the consequences of the existing international division of labour in science, one could be strongly inclined to propose major changes in the pattern of R & D spending in the advanced countries. On the other hand, we must recognize that there would be considerable problems in getting such major changes, both on technical grounds and because of the changes in political priorities which they imply.

In the light of these considerations, we are led to propose that the advanced countries should devote some 5% of their total R & D expenditure in the target year to specific problems of the developing countries. Assuming that advanced countries' GERD/GNP ratios reach 2.5% during the decade, this would represent about 0.13% of their GNP on the average. In the context of a considerably expanding total R&D effort in the advanced countries, this target seems well within the bounds of realistic expectation.

Taken together with the increased efforts of the developing countries themselves and direct aid flows to support these efforts, the target would imply a modest re-orientation of world scientific effort. The total expenditure on R & D for and by the developing countries would be roughly equivalent to present expenditures on space research. This effort would, however, be devoted to a relatively large number of specific objectives.

We reiterate that the feasible target for reorientation of advanced-country effort seems to be well below what would be desirable. In particular, if the coming decade sees a slowdown in the growth of expenditure on military science, there would be a strong argument for a reorientation of effort greater than we have proposed.

SOME IMPLICATIONS FOR THE UNITED NATIONS

A greater allocation of resources to STS in the developing countries, or to problems of specific interest to less-developed countries, could be done either bilaterally or multilaterally. There are certainly considerable advantages in deploying part of the new resources multilaterally, including activities under the auspices of the United Nations. This also applies to the proposal for establishing technology transfer banks. Further study may indicate that an international bank for the transfer of technology within the United Nations system is the best solution. Within the next development decade, the resources of the United Nations system, both in the International Bank and in the United Nations Development programme, are expected to increase considerably. Such increases are already in sight for the next few years. Expanded activities by the United Nations to intensify the application of science and technology in the developing countries should represent a high priority claim on these additional resources. Changes in rules and ways of operation, and possibly also in organization, would be required. No detailed proposals for such changes are made in this part of the World Plan of Action, but it is likely that the present rules of operating only on strict national country requests, and also the rule of not supporting activities carried out within the developed countries, will have to be revised.

"Coupling"—The Problem of Access to World Science and Technology

The proposals in the preceding sections cover efforts required from both the advanced and the developing countries. The success of the programmes as a whole depends on the proper linking of these various efforts, and also on specific action to solve problems of access to foreign technology.

THE "COUPLING" OF SCIENTIFIC ACTIVITIES BETWEEN ADVANCED AND DEVELOPING COUNTRIES

The importance of "coupling" in the science and technology systems has major implications for the manner in which scientific and technological activities are organized in the industrialized countries to aid the developing countries. However well intentioned and however generously financed, such efforts are likely to be largely ineffective unless there is direct personal contact with scientists and technologists in the developing countries themselves. Wherever possible, part or all of the scientific work should actually be carried out in the developing countries. When this is not feasible, generous provision for travel and first-hand acquaintance with the real problems are absolutely essential. Even where these conditions are satisfied, coupling and information flows will remain difficult problems. Great ingenuity and flexibility will be necessary to overcome the communication difficulties, which are much greater in international transfer than in purely internal transfer. Any international or bilateral scheme should have scope for a wealth of independent activities and local variations.

As we have shown, the establishment of a scientific capacity in the developing countries involves building up a range of different scientific and technological organizations. In addition to research laboratories, these will include resource survey organizations, testing facilities, scientific information services and so on. These institutions also can benefit from effective coupling with similar organizations in the industrialized countries.

The leading industrial, university and government laboratories all have an important role to play in carrying out R & D for the developing countries, but it will require a major impulse from central governments and strong financial support to bring about reorientation on the scale required. Extramural grants and contracts on a generous scale should be available to scientific organizations which propose to undertake projects of direct relevance and benefit to the developing countries. Major programmes should be initiated in all the industrialized countries and not just in one or two. There are already quite a number of universities running projects on problems of the developing countries, and this should become the norm in every university. The more far-sighted science-based industrial corporations in Europe, North America and Japan have already started programmes in this field and recognize the urgency of the problem. It should become the normal thing for industrial R & D laboratories to have "Third World" contracts and projects. All these various efforts will have to be linked in with the scientific work going on in the developing countries themselves.

In all this, however, a necessary condition is that scientific work on the problems of the developing countries must achieve the same legitimacy in the eyes of the scientific community as, say, work on atomic programmes or the space programme has today. This will, no doubt, be achieved in part if "development science" is sanctioned by sufficiently large budgetary allocations. But this is only part of the story. A further major factor is that scientists will make a greater contribution to development needs if they know what objectives are relevant both to their own work and to development requirements; and if these objectives are relevant to the criteria for judging scientific work.

This again involves new links between science in advanced and developing countries, particularly in the education of scientific workers. At present, science education trains the young scientist in the methodology of research and introduces him to the body of knowledge in his area of specialization. It is granted that research training should focus on the methodology of fundamental research. Most science education, however, stops at this point. There is a strong case for including in the education of scientists, some training in the analysis of social and economic problems, and in methods of applying scientific expertise to their solution. Science education that stops at fundamental physical science training is not enough; a considerable contribution can be made through multidisciplinary team teaching, where course work is centred around national social issues; for example, the urbanization crisis, agricultural inefficiency, problems of small manufacturing industries, etc. The purpose would be to encourage students to define the social issues in terms meaningful to their own experience, and then apply their own scientific research competence to these issues.

At the same time, efforts will be needed to change the reward and incentive structure for scientists. Career success for a scientist is presently evaluated largely in terms of publications in professional journals, where heavy emphasis is placed on a new motivation of the scientific community by sponsoring publication media of sufficiently high status to be acceptable professionally, yet having a heavy emphasis on development problems.

THE SPECIFIC PROBLEM OF ACCESS TO WORLD TECHNOLOGY

In order to enable the developing countries to make better use of available technologies, the impediments to the transfer of technology must be reduced. In part, of course, these impediments result from the limited capacity to absorb technology. But, there are also problems of access to technology which have to be attacked separately. The problem is to give developing countries access to a wider range of technologies for the development of domestic production. In order to do so, it is essential to break out of the present situation in the world "market" for technology where, by and large, the companies which own technology are only willing to have it exploited in the developing countries if they are able to invest directly.

It is clear that if the developing countries are to make greater use of existing technology, then some means must be found for facilitating its transfer from these enterprises which possess it to those who wish to use it. The success of any venture to promote the transfer of technologies to developing countries depends on the concurrence of enterprises in the advanced countries. The problem is to find a mechanism which will meet the interests of these enterprises as well as the interests of developing countries. Such a mechanism was suggested at the second session of UNCTAD held at New Delhi. In principle, we support the idea of an international technology transfer bank or agency similar to that proposed at New Delhi. This institution would, on the one hand, attempt to cover the risks faced by the companies possessing technologies and, on the other, it would help to reduce foreign-exchange costs of technology purchases. The bank or agency might go further: it might finance research on new technologies designed to meet the specifications of the developing countries.

ECONOMIC REORGANIZATION OF THE DEVELOPING COUNTRIES

While these interrelated objectives which deal specifically with science and technology are the main elements in the strategy for the application of science and technology to development, they will make no sense at all unless the conditions for the application of new technologies are created in the developing countries. This may seem trite but, since it is questionable whether the required conditions exist in the developing countries, the point has to be spelled out.

The essential requirements in the developing countries are:

(a) there has to be a "perceived need" for inputs of science and technology to production; and
(b) it has to be possible to mobilize complementary inputs and to create institutional conditions to get technologies applied.

First, there is the question of what is a "perceived need". The inference of the preceding analysis is that the "need" for science and technology in the developing countries is unlikely to take the form of a commercial demand coming from individual producers. The way production is organized is all against this, and even if it were reorganized, it would take some time before such demands eventuated. This means that the "need" for science and technology will be mainly a function of whether and how far planning bodies (and other organizations responsible for development policy) see a role for science and technology in the development effort. This has two important consequences. First, it implies that there has to be a change in the outlook of the groups responsible for development policies and their implementation. Because of their training, planners and others find it difficult to conceive the role of science and technology in development. Setting up national science councils and the like may help to some extent, but the problem of changing conceptions in development policy usually remains. And there are still very considerable methodological problems about how to relate science and technology policies to development policies, which complicate the whole process. Secondly, the "need" for science and technology may only be perceived if one takes a long-term, perspective view of development because many R & D projects take a long time to give results. This means that the "present need" for R & D will only really be "perceived" if perspective planning methods are used in much greater detail than is normally done.

The second question is about creating the conditions for the application of science and technology to production. At one level, this involves planning for the various factor inputs which are needed to operate new technologies. These may be capital inputs, or skilled technicians, or new kinds of managers, or maintenance workers, etc.

But, more fundamentally, the problem is one of reorganization in the economy. It may, for example, be pointless to undertake programmes on new seeds or new agricultural outputs unless there is land reform or reform of tenure systems. Again the availability of labour-intensive technologies will not in itself result in their use, particularly if relative factor prices are unsuitable. The examples could be multiplied. The point is that the rationale for specific science and technology programmes is directly dependent on the seriousness and success of general development policies. Such reorganization is a slow process and political constraints may make it difficult to achieve.

The World Plan of Action in science and technology, further, cannot fail to take account of another important factor, which itself lies outside the field of science and technology. This is the trade factor. It should be possible for the developing countries to utilize their comparative advantage of abundant labour and lower wage rates by specializing on the production for export of labour-intensive products. Greater access by the developing countries to the markets of the richer countries in respect of labour-intensive products would be no panacea: it may well be that much of the apparent advantage of lower wages would be cancelled out by lower productivity. Effective labour cost might not really be lower in spite of lower wage rates. Hence, freer access by developing countries to markets would have to go hand in hand with rising productivity of labour, including intensive technical assistance towards this end. However, given combined progress in raising productivity and giving freer access, a consistent and concessionary trade policy towards developing countries should be considered as an essential twin pillar of a plan for the wider application of modern technology to developing countries.

The implementation of the policies we have proposed will only make sense and only be justified to the extent that the required reorganizations of the national and international economies are achieved.

We cannot fail to note that together the quantitative targets would imply a vast expansion of science and technology effort for the developing countries over the decade. They would imply, over a slightly longer term, a massive increase in new technologies. When we remember that, at present, there is every evidence that the developing countries are unable to absorb available technologies and to exploit scientific and technical potentials, it is clear that the expanded science and technology programme implies the necessity of major changes in economic organization. This point cannot be too strongly emphasized. Science and technology may be necessary for development; but simultaneously, development efforts are necessary if science and technology are to be used.

*Development, Environment and Technology Assessment**

Ignacy Sachs†

1

Homo faber and sorcerer's apprentice, arrogant lord of nature and victim of his own inventions, man destroys through creating, wringing an exorbitant price from society in return for some of his most spectacular technological achievements, and puts the future in jeopardy while firmly believing that he is safeguarding it. These observations are commonplace in themselves. What is new is the awareness of the irreversibility of certain operations: as the scale of human activities grows ever larger, we are no longer threatened with merely local disturbances, but with imbalances affecting the world as a whole.

Should growth therefore be stopped? The remedy would be worse than the disease. In a world of social inequality, to talk of abolishing growth is tantamount to sentencing the majority of mankind to hopeless penury. It is true that the supporters of this theory pay lip-service to the necessity of giving everyone a fairer slice of the cake. But who can take them seriously when we consider how we have been incapable of correcting the maldistribution of income, either between nations or within them, over a quarter of a century of rapid growth which surely should have made this task easier? Moreover, the apocalypse is not for tomorrow: the situation in regard to non-renewable natural resources certainly does not justify the alarmist views of the prophets of doom. The current fashion for Malthusianism is not the first of its kind since Malthus wrote his famous work, and the earlier ones have been disproved by events. One must not conclude from this that such will always be the case, and take refuge in the kind of "epistemological optimism"[1] espoused by the present incongruous alliance of heads of industry and left-wing technocrats. There can be no doubt that spaceship earth is finite, and that consequently the exponential curves of growth must sooner or later reach a limit. It all depends when, and there is

* This paper first appeared in *Int. Soc. Sci. J.*, Vol. XXV, No. 3 (1973). Translated from French.

† Ignacy Sachs, an economist of Polish origin who studied in Brazil, India and Poland and was the director of the Research Centre on Underdeveloped Economies in Warsaw, has been director of studies at the École Pratique des Hautes Études in Paris since 1968. He has been consultant to international organizations and is the co-author of the chapter on economics in *Main Trends of Research in the Social and Human Sciences*. Vol. I: *Social Sciences* (1970). His other publications include *Patterns of the Public Sector in Underdeveloped Economies* (1964), *Foreign Trade and Economic Development* (1965), and *La Découverte du Tiers-Monde* (1971). He contributed an article on "The logic of development" to *Int. Soc. Sci. J.*, Vol. XXIV, No. 1 (1972).

reason to believe that the date is not after all so close at hand that we must call a halt immediately. On the other hand, it is important to eliminate all wastage as soon as possible and to pay heed to the uses we make of growth.

Awareness of the limits of growth only serves to bring home the urgency of the social problems, the necessity of redressing the wrongs done to the Third World countries, which have long been deprived of their resources or forced to sell them cheaply, and the pressing need to bring technology under control and keep a close rein on the purposes for which it is used.[2] In other words, we must concentrate our attention on the interaction between social and natural processes, between the economics of society and those of nature, emphasizing the crucial role of technology in its broadest sense. In so far as cultures may be defined by their ability to produce goods and services (P) corresponding to the needs they generate, we can adopt Leslie A. White's approximate formula[3]:

$$E \times T \times V - P$$

where $E =$ energy, $T =$ technology, $V =$ environment.

However, this does not settle the problem of the precise role of the environment, on the one hand, and of social organization on the other. Do natural conditions play so compelling a part as to mould individual cultures? Anthropologists rightly agree in rejecting blind geographical determinism. But they cannot wholly escape it, since the harmony established between a culture and its underlying natural background becomes a criterion for appraising the culture, in preference to indices such as the quantity of goods available per head of population.[4]

The very concept of natural resources thus acquires a cultural, and therefore historical,[5] dimension, and the extent to which new resources are brought within the economic system of a society forms a sort of mobile frontier between it and the environment. The environment, in fact, comes to mean everything which remains external to the system under consideration, but which partly determines its performance.[6] If the day ever comes when the economic organization of society truly embraces all the resources of spaceship earth and controls their use, then the environment will have been pushed back to the confines of outer space.

What are the motives and methods for the progressive conquest of the environment?

One thesis which is often put forward relates population, the environment and technological progress. Demographic pressure on the natural environment is claimed to be the basis of technological innovation, which removes restraining factors and makes possible a process of cumulative population growth.[7]

In fact, this relationship appears in a very different light depending on the particular case under review. Historical research has brought to light the destructive effect on the environment of certain technological innovations, resulting in decimation of the population. This was, for instance, the case with the building of roads and railways on embankments in Bengal in the second half of the nineteenth century; the obstruction of the natural drainage system attracted malaria-carrying mosquitoes, with obvious consequences for the local population.[8] Between these two extremes are cases where the interaction between population, environment, technology and social organization has to be spelt out each time anew, on the basis of historical or anthropological research. All this is far removed from the grossly oversimplified hypothesis of the MIT survey entitled *Limits to Growth*.[9]

The futurologist cannot proceed very differently from the historian or the anthropologist. It is his duty to spell out, as far as is possible, the impact of technological options on the interacting spheres of the environment and society, taking into account the fact that damage done to the environment today will affect the possibilities of development in the future.

The kind of approach we are here considering is one based on the evaluation of technological practices as opposed to a mere choice between them, as if such were indeed a mere "matter of choice".

2

We need not dwell too long on the methods used to study the choice of practices. They are usually based on a cost–benefit calculation, which takes into account certain indirect effects and costs entailed by the choice for society.[10] Such a procedure certainly marks a step forward compared with the extremist economic reductionism that brings everything down to a straightforward calculation of profitability at the level of the firm. However, it is to be noted, first, that the decisions of private industry are based in the last analysis on the profit criterion (which confines the applicability of cost–benefit methods to cases where the decision-makers belong to the public sector) and, second, that the cost–benefit method is itself markedly influenced by the economic reductionism that we have just criticized. In fact, not only costs but also effects are reduced to a common denominator and usually calculated in terms of money. It is true that the evaluation methods commonly used in Eastern European countries to evaluate the return on investments partly escape this criticism, but their field of application is much narrower, their purpose being to compare alternative techniques of manufacturing one and the same product. The practice of reducing everything to real or theoretical monetary values carries serious risks. Quantification is sometimes carried out by ingenious but wholly artificial methods, always based on the paradigm of the market phenomenon. Are not attempts made to assess certain environmental impacts by determining the willingness of those affected to pay for a less polluted lake and more breathable air?[11] The artificial character of these calculations is plain, since this impoverished frame of mind is incapable of grasping what lies beyond the bounds of marketing, and remains tied to what has been called the "obsolete market mentality",[12] which reigns unchecked until, in the final analysis, only what is quantifiable is considered to be relevant. Lastly, cost–benefit methods are splendidly malleable in the hands of interest groups: it is enough to interrupt the sequence of effects at such and such a point, to shorten or lengthen the time span as appropriate, or to use a certain procedure for expressing non-monetary costs, to manipulate calculations freely. We know from experience that bad development projects have always been adopted on the basis of apparently flawless and very attractive cost–benefit calculations.[13] Only in delving retrospectively into the circumstances surrounding the decision do we realize the complex interplay of interests involved and the strategies deployed to arrive at the cost–benefit calculations and thus elicit the desired decisions.

It may be pointed out that the abuse of a method is not an argument against its intrinsic merit. The objection is valid, but methods must not be divorced from the context in which they are used and the purposes they serve. Moreover, emphasis on quantities, and their spurious appearance of precision, are means to an ideological end, namely the justification of technocracy.[14]

There cannot, indeed, be a method of evaluation independent of the scale of values chosen as a criterion. These values are often implicit; hence it is necessary first of all to make them explicit. The situations examined are usually marked by the presence of conflicting interests and strategies; it is therefore necessary for the evaluator to state his premises. If there appears to be a social consensus, the task is easier. If the conflicts cannot be resolved, it is appropriate to make an evaluation taking the point of view of the planner and that of the anti-planner in turn. As can be seen, we have come a long way from the confident and alleged neutrality of the cost–benefit approach. It is scarcely surprising that in the circumstances we hesitate to use a weighted scale in order to reduce to a common denominator the many and varied criteria involved.

True enough, once a decision has been taken it is possible to try to deduce the implicit weighting factors and order of priorities that follow from it, but the self-same decision may be the outcome of different combinations of the weights attached to the various criteria. Moreover, to equate the process of decision-making with the choice of a weighting system is to exclude two essential aspects of evaluation, first, the process of spelling out as fully as possible all the future consequences for the environment, science and society of the technological practice evaluated, together with the contexts in which it will be used, and second, the fact of entering into discussion with all the potential agents so as to gain a better understanding—and, where necessary, harmonization—of their attitudes and strategies.[15]

Technology assessment can therefore be seen to go beyond cost–benefit analysis both in width and in depth. It retains the same fundamental procedure, which consists in tracing effects beyond those registered by the market and in taking indirect costs into account. But it carefully avoids seeking a common denominator at all costs and remaining a prisoner of the limitations of the exclusively quantitative approach. Its chief aim is, by means of a systems approach, to encompass all the elements, interests and relationships involved, and on this basis to start an exchange between the agents concerned. A break must be made with the Cartesian method of dividing up the difficulties and tackling them one by one, as if it were certain that the whole added up to no more than the sum of its parts. As Alain Touraine recently recalled, we are no longer in the mechanical world of Descartes—nor, incidentally, are we in the evolutionist world of the nineteenth century. Today our categories of thought and hence of action are different.[16] Technology assessment draws its inspiration essentially from the systems approach. Clearly, such an exercise is meaningful only if the practice evaluated threatens to disrupt existing relations, and if its effects on the environment and on society prove significant. The method outlined above is to be used only to evaluate substantial technological breakthroughs or, by analogy, major industrial and urban developments. It is particularly desirable to have recourse to it where the technology assessed (or the development studied) is likely to spark off consequences for the natural environment or for society that can be reversed only with difficulty or not at all.

3

Recent discussions on the environment conceal an ambiguity. Confusion often arises between the ultimately limited stock of non-renewable resources and the quality of the surroundings in which we must live.

In assessing technology practices it is necessary to draw a clear distinction between these two aspects and to analyse separately their effects on the situation with regard to resources and on the state of our natural surroundings. In addition, of course, their effects on society must be analysed. In this sense, environment and development are seen to be closely connected. The aim is in fact to spell out society's ultimate objectives, making specific mention of the following, closely related, spheres of activity:

- Management of natural resources to avoid ruthless growth likely to mortgage the future.
- Protection of the natural environment to prevent a deterioration of the quality of life in the strict, almost biological, sense of the word.
- Development of productive forces and the establishment of institutions capable of ensuring increased production of goods and services and their fair distribution, which, among other things, implies full employment.
- Development of research directed to the above objectives.

Careless use of non-renewable resources could ultimately jeopardize the future performance of the economy in so far as these resources are really exhaustible and irreplaceable. While the alarmist views of the Club of Rome seem to us to be highly exaggerated, it does not by any means follow that industrialized countries should continue to waste resources as they are doing at present. As far as damage to the environment is concerned, it is nearly always the result of an attitude which subordinates the over-all viewpoint of development to the particular viewpoint of a firm and, more generally, a narrowly economic outlook. Left to itself, business will always try to internalize profits and externalize costs: in other words, it will try to pass on the costs to society. Does this mean that there is an insoluble conflict between protection of the environment and growth?

We by no means believe this to be so. A sound development strategy can perfectly well include an environmental component, which is also a long-term insurance scheme for society; since, moreover, preventive action is usually much less costly than remedial action after the event, by giving some priority to the former one is merely planning the process of development in the most rational way, as seen over a longer time-scale than that appropriate to individual firms. To combine rational management of the environment with economic and social development implies the perfection of suitable technologies, hence the vital importance of scientific and technological policy in all development strategies.

There cannot be a single all-embracing formula, or even several ready-made formulae, to fit the great variety of concrete situations of different societies. This is why we have limited ourselves to distinctions of levels. Each case must be analysed separately, in a concrete manner, and in its historical context. Technological assessment, in the sense here given to it, is possible only by reference to a particular blueprint of civilization.[17] A general observation is nevertheless in order at this point: we must avoid at all costs the sectoral and partial approach that consists in isolating a technical fact, however important, and examining it out of context, when contextual analysis, on the contrary, makes it possible to grasp the whole pattern of relationships between the totality of goals pursued by society, and between its structures and products. Thus, for example, alternative systems of transport and their effects on the environment and on society cannot be assessed without broaching the position occupied by the automobile and petroleum industries in industrialized countries, not to mention the fixed investments already made in the infrastructure as a result of the preponderance of motor transport. From this it must not be concluded that the die is cast once and for all; public transport deserves to be rehabilitated as soon as possible. On the other hand, we must realize that we are not dealing with strictly technical propositions. The demand is for a fundamental change in industrial structures. Paradoxically, it may be inferred from this that some countries of the Third World are perhaps in a better position than the industrialized countries to put truly innovatory solutions into practice in this vital area, inasmuch as they are not yet wholly imprisoned by the consumer patterns of the so-called affluent societies.

4

It is in this light that one should interpret and, where necessary, qualify Barry Commoner's indictment of technology as the primary source of environmental disruption.[18] Whenever a technical alternative more favourable to the environment is put forward, it is necessary to check that the substitute is equivalent to the product in question. Commoner is right when he pleads for a more extensive, less fertilizer- and insecticide-intensive agriculture in the case of the United States. On the other hand, his argument becomes shaky when he urges a return to soap in place of detergent, since the properties of these two products are not identical. Replacing detergents involves a change in the pattern of consumption. Such a change is quite feasible and may prove to be desirable, but the action is no longer in the same category as that of choosing the best method of producing wheat. This leads us to stress that the search for technical solutions to environmental problems must fall within the framework of a consistent, integrated policy covering the following fields:

Research into non-polluting techniques, and measures to encourage reprocessing and the recovery of material from waste matter, as opposed to having recourse to remedial measures: the production–pollution–anti-pollution cycle cannot be allowed to continue escalating as at present.

Control of the obsolescence rate: instead of encouraging, as at present, a rapid turnover of equipment, a premium should be placed on durability and quality, and this does not seem to be beyond the powers of an appropriate tax policy pursuing aims that are the opposite of today's.

Implementation of an industrial location and urbanization policy more in accordance with the requirements of rational resource management and safe-guarding the quality of the environment: the range of possibilities in this field extends from the choice of a particular site to the redistribution of industries on a world scale. Within the framework of an industrial division of labour a more important place should be accorded to the countries of the Third World, which are in a position to adopt an industrial policy for the protection of the environment at a cost substantially lower than for most industrialized countries, and also have abundant non-polluting sources of energy at their disposal; moreover, bulky ocean transport must be restricted, and raw materials produced in the Third World should accordingly be processed on the spot. To this must be added the argument already adduced, that it would be to the advantage of these countries, and wholly within their rights, to seek to turn to good account, domestically, their non-renewable resources of which they have been hitherto dispossessed.

Changes in the pattern of consumption and the organization of the production system with a view to eliminating wastage of resources and energy.

This brings us back to the problem of choosing a "blueprint of civilization", of which the environment, as we have said before, constitutes only one aspect. The terms of this choice are particularly dramatic for the Third World countries, which are running a race against time and cannot allow themselves the luxury of passing through a phase of brazen wastage like that which characterizes industrial capitalism in the developed countries.

In this connection it is important to be quite clear as to what system of values the Third World countries should adopt in themselves assessing the technology which the developed countries have to offer and at the same time in deciding on their own research priorities.

It is by now generally agreed that most industrial techniques of the developed world are ill-adapted to the social and economic circumstances of the Third World. They do not create enough jobs, require considerable outlay in scarce foreign exchange and perpetuate the state of technological dependence of the Third World countries; they also diffuse an alienating pattern of consumption, imported ready-made and accessible only to an élite. Underdevelopment is therefore the result, inasmuch as it consists precisely in the inability to take and implement independent decisions sustained by a continuous flow of innovations of one's own.[19]

Study of the effects of transplanted technologies on the natural environment has only just begun: the picture here is almost equally depressing. For example, in many humid tropical countries, where the specificity of local ecosystems is particularly pronounced, instead of taking this fact into account in the choice of crops and techniques, fruitless and expensive efforts are being made to change these ecosystems so as to make them more like those of temperate climates and thereby adapt them to the use of techniques that have proved their worth in other climates and in other circumstances. This approach must, of course, be totally reversed.

The green revolution, portrayed by some as a panacea to the food problems of the Third World, nonetheless combines both negative features that we have just mentioned. At the social level it polarizes rural communities by favouring a limited group of farming entrepreneurs who have access to water, loans and new techniques. By the force of logic these entrepreneurs are led to mechanization. Following an employment-creation phase, the green revolution will therefore end up by aggravating unemployment. At the ecological level it runs a threefold danger: the elimination of diversity with the risk of increased vulnerability to epidemics; excessive use of water; and ever-increasing recourse to synthetic fertilizers and insecticides. This does not mean, as certain zealots of biological farming would have us believe, that all use of fertilizers and insecticides is inherently bad. There seems to be another possibility open to the Third World countries, both in the social organization of agriculture and in research: in particular, instead of distributing a few miracle strains of wheat, it would perhaps have been better to study the very varied production possibilities of each soil, in collaboration with the farming populations themselves.

5

The foregoing reflections clearly show that, taking the environment into account, there is considerable latitude as regards choice of techniques and also, more basically, as regards criteria for technology assessment. Is it then possible to develop a precise methodology for assessing technology which can be applied indiscriminately in all cases?

We do not think so. Naturally, it is possible to formulate practical rules for the purpose of constructing matrices to register environment impacts or analyse the interlocking interests in play.[20] But to go further would be to slip back into the very straitjacket of cost–benefit analysis and routine from which the method based on technology assessment seeks to break free. Hence, we believe that this method marks an important stage in the development of the applied social sciences. From now on it will be necessary to set up a specific analytical framework for each case, preparatory to studying the interaction between natural and social processes. The necessity of setting up this framework afresh each time, and renewing the analytical tools, may at first glance seem discouraging. However, in addition to guarding against routine work, the ravages of which in the social sciences are too well known to be dwelled upon, it permits rational, informed discussion on the proper role of planning techniques by stressing the institutionalization of social dialogue. If by this means we can gain the necessary experience to exercise social control over technology and science and their uses, we shall have taken a decisive step forward.

Emphasis must be placed on the responsibility of the scientific community to evaluate the probable effects of research results. We therefore endorse the proposal made recently at the twenty-second Pugwash Conference on Science and World Affairs, to earmark 1 to 5% of the budget of each technological research project for the assessment of its consequences.[21]

Notes

[1] The expression is borrowed from Academician B. Kouznetsov, *La Science de l'An* 2000—*Une Philosophie Optimiste de l'Age Atomique* (translated from the Russian), Verviers, 1972.

[2] On this subject see G. Picht, *Réflexions au Bord du Gouffre*, Paris, 1970, and the OECD report *Science, Growth and Society*, 1971.

[2] L. A. White, *The Evolution of Culture*, p. 49, New York, N.Y., 1959.

[4] See, for example, C. Lévi-Strauss, *Race and History*, Paris, 1959. The same author describes as follows the dialectical process which is set up between history and ecology: "Therefore, ethnographical observation does not confront us with the alternative of either a plastic mind passively reflecting the ecology outside, or universal psychic laws unfolding everywhere the same inborn properties, mindless of the ongoing history of each group and the concrete features of its natural and social surroundings. Rather, we witness, and should try to describe, ever changing attempts to compromise between given historical trends and special characteristics of the environment on the one hand, and on the other hand, fundamental psychic requirements which, at each stage, are the outcome of previous ones. As a result, human history and natural ecology become articulated so as to make up a meaningful whole" (C. Lévi-Strauss, 'Structuralism and ecology', *Social Science Information*, Vol. XII, No. 1, 1973).

[5] See L. A. White, *op. cit.*, and also C. O. Saver, *Agricultural Origins and Dispersals*, Bowman Memorial Lectures, New York, N.Y., 1952.

[6] "Not only is the environment something that is outside the system's control but it is also something that determines in part how the system performs" (C. Churchman, *The Systems Approach*, p. 36, New York, N.Y., 1968).

[7] See, for example, E. Boserup, *The Conditions of Agricultural Growth*, London, 1965.

[8] See I. Klein, "Malaria and mortality in Bengal 1840–1921", *Indian Economic and Social History Review*, Vol. IX, No. 2 (1972), pp. 132–60. See also E. Whitcomb, "Development projects and environmental disruption: the case of Uttar Pradesh, India", in: *Political Economy of Environment. Problems of Method*, Paris, 1971, and *The Careless Technology, Ecology and International Development*, New York, N.Y., 1972.

[9] *Limits to Growth*, London, 1972.

[10] See, for example, *Guidelines for Project Evaluation*, Vienna, UNIDO, (United Nations Publication E.72 II.B.11).

[11] For example, a Swedish research worker proposes to calculate the cost of the pollution of a lake by the cost of transport borne by the population in travelling to the nearest unpolluted lake (P. Bohm, "An approach to the problem of estimating demand for public goods", *The Swedish Journal of*

Economics, Vol. 73, No. 1, 1971). The same method is used by N. W. Mansfield, "The estimation of benefits from recreation sites and the provision of a new recreation facility", *Regional Studies*, Vol. V (1971), pp. 55–59 (taken up again in *Cost–Benefit Analysis*, p. 209–18, Harmondsworth, 1972). For criticism of these methods see the contributions of K. W. Kapp, P. Streeten and I. Sachs in *Political Economy of Environment. Problems of Method, op. cit.*

[12] See K. Polanyi, "Our obsolete market mentality", in: G. Dalton (ed.), *Primitive Archaic and Modern Economics: Essays of Karl Polanyi*, pp. 59–67, New York, N.Y., 1968. The same idea is expressed even more forcefully by I. L. McHarg: "The economists, with some conspicuous exceptions, have become the spokesmen for the merchants' creed and in concert they ask with the most barefaced effrontery that we accommodate our values to theirs. Neither love nor compassion, health nor beauty, dignity nor freedom, grace nor delight are true unless they can be priced. If not, they are described as nonprice benefits and relegated to inconsequence, and the economic model proceeds towards its self-fulfilment—which is to say more despoliation. The major criticism of this model is not that it is partial (which is conceded by its strongest advocates), but more that the features which are excluded are among the most important human values, and also the requirements for survival" ("Values, processes and form", in: *The Ecological Conscience—Values for Survival*, p. 26, ed. R. Disch, Englewood Cliffs, N.J., 1970).

[13] A paperback which has become a best-seller in the United States contains many examples drawn from the practices of the Engineering Corps: G. Laycock, *The Diligent Destroyers*, New York, N.Y., 1970.

[14] Some scientists go further than the economist in denouncing the dangers inherent in the abuse of quantities. On this subject see the following passage from Christopher Freeman's article "Malthus with a computer" (*Futures*, Vol. 5, No. 1 (Feb. 1973), p. 12) which is given over to criticism of the report of the Club of Rome: "One of our colleagues in the University, Dr. R. Golub, a physicist, has argued strongly that the MIT approach is inherently dangerous, since it encourages self-delusion in five ways:

By giving the *spurious appearance of precise knowledge* of quantities and relationships which are unknown and in many cases unknowable.

By encouraging the neglect of factors which are difficult to quantify such as policy changes or value changes.

By stimulating gross over-simplification, because of the problem of aggregation and the comparative simplicity of our computers and mathematical techniques.

By encouraging the tendency to treat some features of the model as rigid and immutable.

By making it extremely difficult for the non-numerate or those who do not have access to computers to rebut what are essentially tendentious and rather naive political assumptions.

In short, Golub believes that the method is an attempt to substitute mathematics for knowledge and computation for understanding. See also the excellent study by C. Maestre and K. Pavitt, *Analytical Methods in Government, Science Policy and Evaluation*, Paris, OECD, 1970.

[15] Here we come to so-called strategic planning (see, for example, J. K. Friend and J. C. Yewlett, "Public planning as a multi-organizational process", in *Beyond Local Government Reform; Some Prospects for Evolution in Public Policy Networks*, London, Institute for Operational Research, a unit of the Tavistock Institute of Human Relations, 1971).

[16] A. Touraine, *Proc. Symp. 'Electricité* 2000', 7–9 June 1972, n.d.

[17] I. Sachs, "Environnement et projet de civilisation", *Temps Modernes*, No. 316.

[18] B. Commoner, *The Closing Circle*, New York, N.Y., 1971.

[19] I. Sachs, "L'Indépendance du Tiers Monde: structures de production et structures intellectuelles", in *La Liberté et l'Ordre Social, Rencontres Internationales de Genève* 1969, Neuchâtel, 1969.

[20] For a general presentation of the method based on evaluation of technological practices see the report of the National Academy of Sciences on Technology: *Processes of Assessment and Choice*, Washington, D.C., 1969. See also the annotated bibliography in *Evaluation de l'Environnement*, interim

report of the Groupe Interministériel d'Évaluation de l'Environnement, Paris, Documentation Française, 1973, p. 1.33–4. As regards the "environmental impacts statements" prescribed by law in the United States of America see *Environmental Quality*, Third Annual Report of the Council on Environmental Quality, Washington, D.C., 1972, chap. VII, pp. 221–68.

[21] See M. Thee, 'The scientist's role in society', *Bulletin of Peace Proposals*, No. 4, 1972, pp. 367–70.

CHAPTER IV

The Green Revolution

*The Green Revolution: for Bread and Peace**

NORMAN E. BORLAUG†

CIVILIZATION as it is known today could not have evolved, nor can it survive, without an adequate food supply. Yet food is something that is taken for granted by most world leaders despite the fact that more than half of the population of the world is hungry. Man seems to insist on ignoring the lessons available from history.

Man's survival from the time of Adam and Eve until the invention of agriculture must have been precarious because of his inability to ensure his food supply. During the long, obscure, dimly defined prehistoric period when man lived as a wandering hunter and food gatherer, frequent food shortages must have prevented the development of village civilizations. Under these conditions the growth of human population was also automatically limited by the limitations of food supplies.

In the misty, hazy past, as the Mesolithic Age gave way to the Neolithic, there suddenly appeared in widely separated geographic areas the most highly successful group of inventors and revolutionaries that the world has ever known. This group of Neolithic men and women, and in all probability largely the latter, domesticated all the major cereals, legumes and root crops as well as all of the most important animals that to this day remain man's principal source of food.

Apparently, 9000 years ago, in the foothills of the Zagros Mountains, man had already become both agriculturist and animal husbandry-man, which, in turn, soon led to the specialization of labor and the development of village life. Similar discoveries and developments elsewhere soon laid the groundwork from which all modern agriculture and animal industry and, indeed, all of the world's contributions stem. We know none of these benefactors of mankind by name. In fact, it has only been within the past century, and especially within the last 15 years—since the development of the effective radiocarbon dating system—that we have begun even vaguely to understand the timing of these epochal events which have shaped the world's destiny.

The invention of agriculture, however, did not permanently emancipate man from the fear of food shortages, hunger and famine. Even in prehistoric times population growth often must have threatened or exceeded man's ability to produce enough food. Then, when droughts or outbreaks of diseases and insect pests ravaged crops, famine resulted.

* Laureate address presented to Nobel Foundation upon receipt of Nobel Peace Prize, awarded in October 1970.

† Director, International Wheat Research and Production Program, International Maize and Wheat Improvement Center, Mexico City.

That such catastrophes occurred periodically in ancient times is amply clear from numerous biblical references. Thus, the Lord said: "I have smitten you with blasting and mildew" (Amos 4:9); "The seed is rotten under their clods, the garners are laid desolate, the barns are broken down; for the corn is withered. . . . The beasts of the fields cry also unto thee: for the rivers of waters are dried up, and the fire hath devoured the pastures of the wilderness' (Joel 1:17, 20).

Plant diseases, drought, desolation, despair—were recurrent catastrophes during the ages. And the ancient remedies: supplications to supernatural spirits or Gods. And yet, the concept of the "ever-normal granary" appeared in elementary form, as is clear from Pharaoh's dreams, and Joseph's interpretation of imminent famine and his preparation for it, as indicated by this quotation from Genesis: ". . . And the seven years of dearth began to come, according as Joseph had said: and the dearth was in all lands; but in all the land of Egypt there was bread . . ." (Genesis 41:54). For his time, Joseph was wise, with the help of his God.

But today we should be far wiser; with the help of our gods and our science we must not only increase our food supplies but also insure them against biological and physical catastrophes by international efforts to provide international granaries of reserve food for use in case of need. And these food reserves must be made available to all who need them—and before famine strikes, not afterwards. Man can and must prevent the tragedy of famine in the future instead of merely trying with pious regret to salvage the human wreckage of the famine, as he has so often done in the past. We will be guilty of criminal negligence, without extenuation, if we permit future famines. Humanity cannot tolerate that guilt.

Alfred Nobel was also very conscious of the importance of food, for he once wrote: "I would rather take care of the stomachs of the living than the glory of the departed in the form of monuments."

The destiny of world civilization depends upon providing a decent standard of living for all mankind. The guiding principles of the recipient of the 1969 Nobel Peace Prize, the International Labor Organization, are expressed in its charter in the words, "Universal and lasting peace can be established only if it is based upon social justice. If you desire peace, cultivate justice." This is magnificent; no one can disagree with this lofty principle.

Food, A Moral Right

Almost certainly, however, the first essential component of social justice is adequate food for all mankind. Food is the moral right of all who are born into this world. Yet today 50% of the world's population goes hungry. Without food man at most can live but a few weeks; without it, all other components of social justice are meaningless. Therefore I feel that the forementioned guiding principle must be modified to read: If you desire peace, cultivate justice, but at the same time, cultivate the fields to produce more bread; otherwise there will be no peace.

The recognition that hunger and social strife are linked is not new, for it is evidenced by the Old Testament passage "... and it shall come to pass that when they shall be hungry, they shall fret themselves and curse their King and their God ..." (Isaiah 8:21).

It is a sad fact that on this earth at this late date there are still two worlds, "the privileged world" and "the forgotten world". The privileged world consists of the affluent, developed nations, comprising 25 to 30% of the world population, in which most of the people live in a luxury never before experienced by man outside of the Garden of Eden. The forgotten world is made up primarily of the developing nations, where most of the people, comprising more than 50% of the total world population, live in poverty, with hunger as a constant companion and fear of famine a continual menace.

When the Nobel Peace Prize Committee designated me the recipient of the 1970 award for my contribution to the Green Revolution, they were in effect, I believe, selecting an individual to symbolize the vital role of agriculture and food production in a world that is hungry, both for bread and for peace. I am but one member of a vast team made up of many organizations, officials, thousands of scientists, and millions of farmers—mostly small and humble—who for many years have been fighting a quiet, oftentimes losing war on the food-production front.

During the past 3 years spectacular progress has been made in increasing wheat, rice and maize production in several of the most populous developing countries of southern Asia, where widespread famine appeared inevitable only 5 years ago. Most of the increase in production has resulted from increased yields of grain per hectare, a particularly important development because there is little possibility of expanding the cultivated area in the densely populated areas of Asia.

The term "Green Revolution" has been used by the popular press to describe the spectacular increase in cereal-grain production during the past 3 years. Perhaps the term Green Revolution, as commonly used, is premature, too optimistic, or too broad in scope. Too often it seems to convey the impression of a general revolution in yields per hectare and in total production of all crops throughout vast areas comprising many countries. Sometimes it also implies that all farmers are uniformly benefited by the breakthrough in production.

These implications both oversimplify and distort the facts. The only crops which have been appreciably affected up to the present time are wheat, rice and maize. Yields of other important cereals, such as sorghums, millets and barley, have been only slightly affected; nor has there been any appreciable increase in yield or production of the pulse or legume crops, which are essential in the diets of cereal-consuming populations. Moreover, it must be emphasized that thus far the great increase in production has been in irrigated areas. Nor have all cereal farmers in the irrigated areas adopted and benefited from the use of the new seed and the new technology. Nevertheless, the number of farmers, small as well as large, who are adopting the new seeds and new technology is increasing very rapidly, and the increase in numbers during the past 3 years has been phenomenal. Cereal production in the rain-fed areas still remains relatively unaffected by the impact of the Green

Revolution, but significant change and progress are now becoming evident in several countries.

Our Daily Bread

Despite these qualifications, however, tremendous progress has been made in increasing cereal production in India, Pakistan, and the Philippines during the past 3 years. Other countries that are beginning to show significant increases in production include Afghanistan, Ceylon, Indonesia, Iran, Kenya, Malaya, Morocco, Thailand, Tunisia and Turkey.

In the affluent, industrialized nations giant surpluses of wheat, maize and sorghum are commonplace; cattle, swine, and poultry are fed and fattened on cereal grains; meat, milk, eggs, fruits and vegetables are within the economic reach of most of the population; well-balanced diets are more or less automatically achieved, and cereal products constitute only a modest portion of the "daily bread". Consequently, most of the people in such societies have difficulty in comprehending and appreciating the vital significance of providing high-yielding strains of wheat, rice, maize, sorghum and millet for the people of the developing nations. Understandably, then, the majority of the urbanites in the industrialized nations have forgotten the significance of the words they learned as youngsters, "Give us this day our daily bread". They know the food comes from the supermarket, but only a few see beyond to the necessary investments, the toil, struggle, and frustrations on the farms and ranches that provide their daily bread. Since the urbanites have lost their contact with the soil, they take food for granted and fail to appreciate the tremendous efficiency of their farmers and ranchers, who, although constituting only 5% of the labor force in a country such as the United States, produce more than enough food for their own nation.

Contrasting sharply, in the developing countries, represented by India, Pakistan and most of the countries in Asia and Africa, 70 to 80% of the population is engaged in agriculture, mostly at the subsistence level. The land is tired, worn out, depleted of plant nutrients and often eroded; crop yields have been low, near starvation level and stagnant for centuries. Hunger prevails and survival depends largely upon the annual success or failure of the cereal crops.

The significance and magnitude of the impact of the so-called Green Revolution is best illustrated by changes in cereal production in India, Pakistan and the Philippines. In both India and Pakistan the rapid increase in yields per hectare of wheat has been the major thrust of the Green Revolution. Increases in rice yield also have played a major role in West Pakistan but hitherto only a minor role in India. Increases in maize production have played a modest but significant role in expanded cereal production in both India and Pakistan; and increases in rice yields and production have been largely responsible for the change in cereal production up to now in the Philippines, Ceylon and Indonesia.

The Green Revolution in India and Pakistan, which is still largely the result of a breakthrough in wheat production, is neither a stroke of luck nor an accident of nature. Its success is based on sound research, the importance of which is not self-evident at first glance. For, behind the scenes, halfway around the world in Mexico, were two decades of aggressive research on wheat that not only enabled Mexico to become self-sufficient with respect to wheat production but also paved the way to rapid increase in its production in other countries. It was in Mexico that the high-yielding, Mexican dwarf varieties were designed, bred and developed. There, also, was developed the new production technology which permits these varieties, when properly cultivated, to express their high genetic grain-yield potential—in general, double or triple that of the best yielders among older, tall-strawed varieties.

Principal Catalyst

There are no miracles in agricultural production. Nor is there such a thing as miracle variety of wheat, rice or maize which can serve as an elixir to cure all ills of a stagnant, traditional agriculture. Nevertheless, it is the Mexican dwarf wheat varieties, and their more recent Indian and Pakistani derivatives, that have been the principal catalyst in triggering off the Green Revolution. It is the unusual breadth of adaptation, combined with high genetic yield potential, short straw, a strong responsiveness and high efficiency in the use of heavy doses of fertilizers, and a broad spectrum of disease resistance that has made the Mexican dwarf varieties the powerful catalyst that they have become in launching the Green Revolution. They have caught the farmers' fancy, and during the 1969–70 crop season 55% of the 6 million hectares down to wheat in Pakistan and 35% of the 14 million hectares in India were sown to Mexican varieties or their derivatives.

This rapid increase in wheat production was not based solely on the use of Mexican dwarf varieties; it involved the transfer from Mexico to Pakistan and India of a whole new production technology that enables these varieties to attain their yield potential. Perhaps 75% of the results of research done in Mexico in developing the package of recommended cultural practices, including fertilizer recommendations, were directly applicable in Pakistan and India. As concerns the remaining 25% the excellent adaptive research done in India and Pakistan by Indian and Pakistani scientists, while the imported seed was being multiplied, provided the necessary information for modifying the Mexican procedures to suit Pakistani and Indian conditions more precisely.

Equally as important as the transfer of the new seed and new technology from Mexico to India and Pakistan was the introduction from Mexico of a crop-production campaign strategy. This strategy harnessed the high grain-yield potential of the new seed and new technology to sound governmental economic policy which would assure the farmer a fair price for his grain, the availability of the necessary inputs—seed, fertilizers, insecticides, weed killers and machinery—and the credit with which to buy them. Collectively these inputs and strategy became the base from which the Green Revolution evolved.

Never before in the history of agriculture has a transplantation of high-yielding varieties coupled with an entirely new technology and strategy been achieved on such a massive scale, in so short a period of time, and with such greatsuccess. The success of this transplantation is an event of both great scientific and social significance. Its success depended upon good organization of the production program combined with skilful execution by courageous and experienced scientific leaders.

Experimentation with dwarf Mexican varieties was initiated in both India and Pakistan in 1963 and continued in 1964. Results in both countries were highly promising. Consequently in 1965, 350 and 250 tons of seed of the Mexican dwarf wheat varieties were imported into Pakistan and India, respectively, for wide-scale testing on farms. Again, the results were highly promising, and India reacted by importing 18,000 tons during 1966. A year later Pakistan imported 42,000 tons. With these imports, the revolution in wheat production got underway in both countries. It was the first time in history that such huge quantities of seed had been imported from distant lands and grown successfully in their new home. These imports saved from 3 to 5 years' time in reaping the benefits from the Green Revolution.

During the past 3 years wheat production has risen spectacularly in both countries. Using as a base the pre-Green Revolution crop year 1964–5, which produced an all-time record harvest in both countries, the production in Pakistan increased from the 1965 base figure of 4.6 millions tons to 6.7, 7.2 and 8.4 millions of tons, respectively, in 1968, 1969 and 1970. West Pakistan became self-sufficient in wheat production for the first time in the 1968 harvest season, 2 years ahead of our predictions. Indian wheat production has risen from the 1964–5 pre-Green Revolution record crop of 12.3 million tons to 16.5, 18.7 and 20.0 million tons during 1968, 1969 and 1970 harvests, respectively. India is approaching self-sufficiency and probably would have attained it by now if rice production had risen more rapidly, because, with a continuing shortage of rice, considerable wheat is being substituted for it.

The introduction into West Pakistan of the high-yielding dwarf rice variety IR8, developed by the International Rice Research Institute (IRRI) in the Philippines, together with the new technology that makes it highly productive, has also resulted in phenomenal increases in yield and production during the past 2 years. Unfortunately, this variety has been less well adapted to climatic conditions in the monsoon areas of India and in East Pakistan and therefore has had only a modest and occasional impact there. Newer varieties which are now being multiplied promise to correct this situation.

The revolution in wheat and rice production in India and Pakistan has not only greatly increased food production, but it also has had many indirect effects on both the farmer and the economy. It is estimated that Indian and Pakistani farmers who are cultivating the new Mexican dwarf wheat varieties under the recommended management practices have increased their net income from \$37 per hectare with the local varieties to \$162 with the dwarf Mexican varieties. During the past three harvests a total of \$1.4 billion and \$640 million have been added to the GNP of India and Pakistan, respectively, from the increase in wheat production above the record 1965 base. The injection of this large increase in purchasing power into the economies has had many effects. Large numbers of tubewells are being sunk by farmers in both India and Pakistan in order to expand the irrigated area and improve the control of irrigation water. It is estimated that a total of 70,000 private tubewells were sunk during the 1969–70 crop season in India, which brings about 1.4 million hectares of additional land under controlled irrigation, thereby greatly expanding the food production potential. It is estimated that at present less than half of the irrigation potential of India has been developed.

Fertilizer Consumption

If the high-yielding dwarf wheat and rice varieties are the catalysts that have ignited the Green Revolution, then chemical fertilizer is the fuel that has powered its forward thrust. The responsiveness of the high-yielding varieties has greatly increased fertilizer consumption. The new varieties not only respond to much heavier dosages of fertilizer than the old ones but are also much more efficient in its use. The old tall-strawed varieties would produce only 10 kilos of additional grain for each kilo of nitrogen applied, while the new varieties can produce 20 to 25 kilos or more of additional grain per kilo of nitrogen applied.

Consumption of nitrogen fertilizer in India has increased from 58,000 metric tons of nutrients in 1950–1, to 438,000 and 1.2 million metric tons in 1964–5 and 1969–70 crop cycles, respectively; and about 60% of this amount was produced domestically. Phosphate consumption is approximately half that of nitrogen. A large part of the fertilizer currently being used is for wheat. The targeted consumption and domestic production needs of nitrogen for 1973–4 are 3 million and 2.5 million metric tons, respectively, a fantastic threefold increase in consumption and a fivefold increase in production. These fertilizer targets must be attained if the targeted production of 129 million metric tons of cereal is to be realized.

Mechanization of agriculture is rapidly following the breakthrough in wheat production. Prior to the first big wheat crop in 1968, unsold tractors accumulated at the two factories then in production; at present, prospective purchasers must make written application for them and wait 1 or 2 years for delivery. Although five factories with an output of 18,000 units per year are now producing tractors, 35,000 units were imported in 1969–70.

The traditional method of threshing by treading out the grain with bullocks, followed by winnowing, is now inadequate for threshing the increased volume of wheat before the onset of the monsoon rains. Consequently, tens of thousands of small threshing machines have been produced and sold by hundreds of small village machine shops during the past 3 years, thus avoiding the loss of much of the crop after harvest and also providing additional employment in many new, small-village industries.

Moreover, mechanization has had another very important indirect effect on the intensification of cereal production. When small mechanical threshers replace bullocks for threshing, the bullocks are released for use in the timely preparation of the land for the next (summer) crop. This need for timely preparation of land is also one of the main reasons for the surge in demand for tractors.

Before the adoption of the new wheat and rice varieties, in combination with heavy applications of chemical fertilizer, the time of sowing was relatively unimportant because yields were limited primarily by low level of available plant nutrients. Most farmers would expect to harvest about 1 metric ton of wheat during the winter (rabi) season and about 1.5 metric tons of rice during the summer (kharif season), or a total of 2.5 metric tons of grain per hectare per year. But by using the high-yielding varieties, fertilizing heavily, sowing at the right time and managing the fields properly, the same farmer can now harvest 5 tons of wheat and 7 tons of rice per hectare, as contrasted with 2.5 tons which he obtained with the old varieties and methods.

If plantings are not done at the optimum time, however, the yield of wheat may drop to 3 tons and that of rice to 4 tons per hectare, a total production of 7 tons per year instead of the 12 tons when all operations are proper and timely. A few of the most progressive farmers now use triple cropping, involving wheat–mung beans–rice, wheat–rice–potato, or three consecutive crops of rice during the same year. By increasing the intensity of cropping, both food-production potential and employment are increased. Yields must then be calculated on the basis of kilos per hectare per year rather than on the basis of kilos per hectare per crop.

The increased mechanization in cereal production has tended thus far to increase rather than decrease the employment opportunities for labor, and above all it has helped to reduce drudgery and increase the efficiency of human energy, speaking especially of India.

Millions of farmers who have successfully grown the new wheat, rice and maize varieties have greatly increased their income. And this has stimulated the rapid growth of agro-industry by increasing the demand for fertilizers, pumps, machinery and other materials and services.

Farmers in many villages are investing in better storage facilities. In some locations brick houses are beginning to replace those made of rammed earth. More electricity is being used to light the houses and to drive the motors on the wells. There also has been a rapid increase in demand for consumer goods. The purchase of transistors and radios for use in the villages has increased rapidly, and thereby the government for the first time can effectively reach the remote villages with educational programs. Sewing machines, bicycles, motor-scooters and motor-cycles are coming to the villages and truck and bus service between villages is improving.

The Green Revolution has forced the Indian Government to improve many of its public services. Although there was an extreme shortage of regular storage space for the first record-breaking wheat crop in 1968, the Government improvized satisfactorily and very little grain was lost. During the past 2 years, stimulated in part by criticism by farmers and the press, warehouse capacity has been expanded greatly to provide adequate storage for the increasing grain production. The villages are demanding better roads, better public transportation and better schools; and they are beginning to get them. Thus the divorce between intellect and labor, which the great Indian leader Mahatma Gandhi, over 40 years ago, regarded as the bane of India's agriculture, is coming to an end.

The changes wrought by the Green Revolution, which I have illustrated by the vast improvement of wheat production in India, have had similar effects in West Pakistan, Ceylon, the Philippines and Thailand, although the effects in different countries were produced by changes in different crops or combinations of crops.

Although the contributions of the Green Revolution to increased food production are considerable and highly significant, they are none the less modest in comparison with the magnitude of present global needs. The greatest obvious achievement is the rapid increase in cereal production during the past 3 years, and the generation of a climate of confidence in the developing nations with regard to their capabilities of achieving food self-sufficiency. Perhaps even more significant, however, is the change in organizations and attitudes which have accompanied the increases in cereal production.

Indian Program

The All-India Coordinated Wheat Improvement Program, which is largely responsible for the wheat revolution in India, has developed one of the most extensive and widely diversified wheat research programs in the world. Its success has generated confidence, a sense of purpose and determination. The current agronomic research on wheat in India equals the best in the world. The breeding program is huge, diversified and aggressive; already it has produced several varieties which surpass those originally introduced from Mexico in 1965. The first group of new Indian varieties, already in extensive commercial production, were derived from selections made in India from partially selected materials received from Mexico. A second group of varieties, now being multiplied, are selections from crosses made in India between Indian and Mexican varieties. The rapidity of creation and distribution of those new varieties has already diversified the type of resistance to diseases and therefore minimizes the menace of destructive disease epidemics if and when changes occur in parasitic races of the pathogens.

Contrary to a widespread and erroneous opinion, the original dwarf wheats imported from Mexico definitely carried a wider spectrum of disease resistance than the local Indian types that they replaced. But the newer Indian varieties are even better in resistance and of a different genetic type than the original introductions. This greater diversity reduces the danger from disease epidemics, but cannot completely eliminate the dangers, as has become vividly evident from the unexpected and destructive epidemic of southern leaf blight of maize over vast areas of the United States during the summer of 1970.

The only protection against such epidemics, in all countries, is through resistant varieties developed by an intelligent, persistent and diversified breeding program, such as that being currently carried on in India, coupled with a broad disease-surveillance system and a sound plant pathology program to support the breeding program. From such a program a constant flow of new high-yielding, disease-resistant varieties can be developed to checkmate any important changes in the pathogens. The Indian program is also developing competence in research on the biochemical, industrial and nutritional properties of wheat.

Perhaps the most important contribution of all is that the methods and tactics used so successfully in making the production breakthrough in wheat—first in Mexico and now in India and Pakistan—can serve as a model for production programs with many other crops and in many other countries.

West Pakistan has already used the wheat model to revolutionize its rice production. Although the Indian rice program has not yet achieved a nationwide breakthrough in production, rapid progress is now being made in several areas and it seems probable that the area sown to the new seed and technology will be large enough to produce a strong impact on national production within another year. Varieties and new technology are also available for launching effective campaigns to increase the production of sorghum, millet, barley, soybeans and cotton in many developing countries of Asia, Africa and Latin America.

What is still needed is the will and commitment of governments to support national production campaigns, both politically and financially, and the services of a few competent and dedicated agricultural scientists as leaders.

The quality of scientific leadership is certainly a vital factor in the success of any production campaign. It is deplorable, but true that many agricultural scientists in some advanced countries have renounced their allegiance to agriculture, for reasons of expediency and presumed prestige. And some institutions have furnished them a curtain behind which to hide. Some educational and research institutions have even restricted the amount of basic research that can be done under the aegis of its agricultural departments, however basic these researches may be to progress in increasing and insuring food production. Let the individuals live with their own motivations; let them serve science and themselves if they wish. But the institutions have the moral obligation to serve agriculture and society also; and to discharge that obligation honorably, they must try to help educate scientists and scientific leaders whose primary motivation is to serve humanity.

I want to reiterate emphatically that there now are available materials and techniques of great potential value for expanding the Green Revolution into additional fields of agriculture. But to convert these potential values into actual values requires scientific and organizational leadership. Where are those leaders? Where are the leaders who have the necessary scientific competence, the vision, the common sense, the social consciousness, the qualities of leadership, and the persistent determination to convert the potential benefits into real benefits for mankind in general and for the hungry in particular? There are not enough of them now; therefore we must try to identify and develop them in our educational systems and we must utilize them in our campaigns for food production. We need them, and need them badly, for it is tragic to let potential values languish for want of leadership in capitalizing the potential. This is not a theory; this is reality, as illustrated by the fact that the leadership has been the determining factor in the relative success of parallel but different crop production programs within the same country.

But let no one think that we can relax our efforts in research. All successful action programs must be preceded and accompanied by research. It has been pointed out that the rapid change in wheat production in both India and Pakistan was in part made possible by two decades of research in Mexico. How did this come about?

In 1943, several years before the establishment of the Food and Agricultural Organization (FAO) of the United Nations, a cooperative agricultural research and training program was launched in Mexico. This was a pioneer cooperative project between the Mexican Ministry of Agriculture and the Rockefeller Foundation, initiated at the request of the Mexican Government for assistance in increasing the production of maize, wheat and beans.

At that time Mexico was importing more than 50% of the wheat that it consumed, as well as a considerable percentage of its maize. Wheat yields were low and static with a national average yield of 750 kilos per hectare, even though most of the wheat was grown on irrigated land. The situation was very similar to that in India and Pakistan before the recent advent of the Green Revolution. Mexican soils were impoverished and chemical fertilizer virtually unknown. Mexico's need was urgent and so a simple research program was started to increase production. The philosophy of the Rockefeller Foundation was "to help Mexico to help itself" in solving its food-production problems, and in the process work itself out of a job. I have had the privilege and good fortune to have been associated with the wheat program almost from the beginning, and have remained a part of it for the past 26 years. From the outset all factors limiting wheat production were studied; consequently there were interdisciplinary researches between genetics and plant breeding, agronomy, soil fertility, plant pathology and entomology. Cereal chemistry and biochemistry were added later.

Concentrated Attack

After preliminary work in 1943, plant breeders, soil scientists, plant pathologists and entomologists, working as a team, began a concentrated attack on the various aspects of wheat production in 1944.

An in-service (intern) training component was added to the research program to train a new generation of Mexican scientists—while they were assisting with the development of the research program. Provision was also made for fellowships to enable the most promising of these young scientists to study abroad for advanced degrees, hopefully in preparation for positions of leadership in Mexican agriculture.

Research from the outset was production-oriented and restricted to that which was relevant to increasing wheat production. Researches in pursuit of irrelevant academic butterflies were discouraged, both because of the acute shortage of scientific manpower and because of the need to have data and materials available as soon as possible for use in the production program.

To accelerate progress in varietal development, two generations of all segregating materials were grown each year. One generation was sown close to sea level in Sonora at 28 degrees north latitude in the fall when the days were progressively shorter; the second was sown near Toluca, at 18 degrees latitude and 2500 meters above sea level during the summer when days were progressively longer. Through the use of this technique we developed high-yielding, day-length insensitive varieties with a wide range of ecologic adaptation and a broad spectrum of disease resistance—a new combination of uniquely valuable characters in wheat varieties.

These characters were valuable in increasing wheat production in Mexico and neighboring countries, but were to prove even more valuable 20 years later when the Mexican varieties were introduced into Pakistan and India. Without this combination of characters the successful transplantation of the Mexican varieties into Pakistan and India would have been impossible, and the advent of the Green Revolution would almost certainly have been delayed many years.

In Mexico, as soon as significant improvements were made by research, whether in varieties, fertilizer recommendations or cultural practices, they were taken to farms and incorporated into the production programs. We never waited for perfection in varieties or methods but used the best available each year and modified them as further improvement came to hand. This simple principle is too often disregarded by scientific perfectionists who spend a lifetime searching for the unattainable in biological perfection, and consequently during a lifetime of frustration contribute nothing to increasing food production.

Farm demonstrations of new varieties and technology were made by the research scientists who had developed them. Indeed, the revolution in wheat production in Mexico was accomplished before the extension service came into being. This forced the research scientists themselves to consider the obstacles to production that confronted the farmers. The same philosophy and tactics were used effectively to bring researchers in contact with the farmers' problems in the early years of the wheat-improvement programs in India and West Pakistan. Later, however, the extension services were brought into the production programs in both countries.

The Quiet Revolution

Mexican wheat yields began to climb by 1948 and have continued their upward trend to the present time. During the past 26 years the national average has risen from 750 kilos per hectare to only slightly less than 3000 kilos during the past harvest, approximately a fourfold increase. During the same period total production has increased sevenfold. Mexico became self-sufficient in wheat production for the first time in 1956, and has remained self-sufficient since. This "Quiet Revolution" in wheat production in Mexico became the progenitor of the Green Revolution in India and Pakistan a decade later.

As the use of fertilizer increased and yields climbed to 4500 kilos per hectare, lodging (falling over of the plant) began to limit further increases in yields. A search was therefore made among wheats from different areas of the world to locate a suitable source of genetic dwarfness to overcome this barrier. Norin 10, an extremely dwarf wheat from Japan, proved to be a suitable source. Through a series of crosses and recrosses, begun in 1954, dwarfness was incorporated into the superior, new-combination Mexican types, finally giving rise to a group of so-called dwarf Mexican wheat varieties. With this new development, the potential yield of the new varieties, under ideal conditions, increased from the previous high of 4500 kilos per hectare to 9000 kilos per hectare. The dwarf Mexican wheats were first distributed in Mexico in 1961, and the best farmers began to harvest 5, 6, 7 and even 8 tons or more per hectare; and within 7 years the national average yields doubled. It was these same dwarf Mexican wheats from the Quiet Revolution that served as catalysts to trigger off the Green Revolution in India and Pakistan.

From the outset the Mexican Agricultural Program was watched with interest by many other countries. As progress became evident, the Rockefeller Foundation was besieged by requests from many countries for assistance in agricultural improvement programs. The Cooperative Mexican Agricultural Program had become a model. The Colombian Agricultural Program, devoted largely to maize, wheat, potatoes, forage and livestock, was established in 1950. Similarly, the Cooperative Chilean Agricultural Program was established in 1955 to work on wheat and forage. The Cooperative Indian Agricultural Program was established in 1956 to improve maize, sorghum and millet production and to assist in the development of postgraduate agricultural education. Each of these programs subsequently played an important role in improving agricultural production and education in different parts of the world.

Meanwhile, back in Mexico, the program that had originally been confined to maize, wheat and beans, and soon thereafter potatoes, was expanded to include many other crops. Larger numbers of young Mexican scientists were added to the research and training programs. Progress in research was generally good, and the training program also bore fruit. Between the years 1943–63 a total of 550 interns participated in the overall agricultural research and training programs, of whom 200 received a Master of Science degree and 30 the Doctor of Philosophy degree while on fellowships for study abroad. With this corps of trained scientists a new National Institute of Agricultural Investigations was born in 1961. The Rockefeller Foundation "had worked itself out of a job", which was one of its original objectives.

The Mexican experience indicated that one of the greatest obstacles to the improvement of agriculture in the developing countries is the scarcity of trained people. This experience indicated clearly that training is a slow process. Where no corps of trained scientists exist, as was the case in Mexico 27 years ago and remains the case in many countries of Asia, Africa and Latin America today, it requires 18 to 25 years to develop enough competent research scientists and educators to meet a country's needs.

The International Centers

So great is the urgency of the food shortage in many underdeveloped and emerging countries that there is not enough time to develop an adequate corps of scientists before attacking food production problems. A short-cut and organizational change had to be invented to meet the needs. And so was born the first truly international research and training institute—the International Rice Research Institute (IRRI)—at Los Banos, the Philippines, in 1960, to work exclusively on the regionally all-important but too-long neglected rice crop. The Institute was jointly financed by the Ford and Rockefeller Foundations in collaboration with the government of the Philippines.

The research activities on wheat, maize and potatoes in Mexico were informally internationalized in 1959 and organized as a second international center in 1963. This International Center for Maize and Wheat Improvement (CIMMYT) is supported also by the Ford and Rockefeller Foundations, in collaboration with the government of Mexico. More recently, additional financial support has been provided by the U.S. Agency for International Development (AID), United Nations Development Program (UNDP) and the Inter-American Development Bank (BID).

A third center, the International Center of Tropical Agriculture (CIAT), in Columbia, and the International Institute of Tropical Agriculture (IITA) in Nigeria, the most recent, have been established to study problems and stimulate production of certain tropical crops and animal species, as well as to help train scientific specialists. CIAT is financed by the Ford, Rockefeller, and W. K. Kellogg Foundations in co-operation with the government of Colombia. The Ford and Rockefeller Foundations, and the Canadian International Development Agency (CIDA) are supporting IITA, in collaboration with the government of Nigeria.

These four international institutes represent a significant but modest start towards the construction of a worldwide network of international, national and local research and training centers. This network will help solve problems and disseminate the benefits of science to all mankind in the shortest possible time and at minimum costs.

The impact of such integrated approach is already evident in the Green Revolution. New varieties and the new technologies that make them highly productive have been the thrust behind the Green Revolution. In the Philippines, Ceylon, Malaysia and West Pakistan it was IR8 rice, developed at the International Rice Research Institute. The dwarf Mexican wheats, partly produced by CIMMYT, have provided the thrust in India and Pakistan and this is now spreading to Turkey, Afghanistan, Iran, Morocco and Tunisia. Contributing equally, or perhaps even more, to the evolution of the Green Revolution was the talented supporting leadership that has been provided by these centers to the national programs through temporary assignments of mature scientists skilled in organizing crop-production programs to assist in the development of the national production campaigns.

The international centers were developed to supplement national agricultural research, production and training programs, not to replace them. The centers are but one link in the worldwide network of organizations attacking basic food-crop-production problems on a worldwide, regional, national and local level. The backbone of this network is now and must continue to be the national programs. These must be given greater financial support and strengthened staff-wise to meet the challenge of rapidly expanding food needs for the future.

The international centers, however, are in a unique position to assist the national programs. They are independent, nonpolitical international organizations, which, although originally funded by private foundations, now receive support from many diverse sources. Their scientific staffs are also international and comprise outstanding scientists representing the various scientific disciplines affecting crop production. Included on their staff are a number of crop-production experts who have the scientific competence and broad experience to assist national agencies in organizing and launching crop production programs.

The centers collaborate not only with the national agencies from many different countries, but also with other international organizations such as FAO, UNDP and international development banks. Each year the centers have been collaborating with an increasing number of countries of all political spectra.

I am convinced that the international agricultural research institutes are developing a bond of understanding among nations, based upon the common need for increasing food production. We must all strive to strengthen this bond in the spirit of Alfred Nobel: "... to promote brotherhood among the nations."

The international centers are uniquely equipped to do fundamental, longtime researches of worldwide importance. For example, the opportunity for plant breeders, pathologists and entomologists to operate on a worldwide basis permits them to develop well-conceived, diverse gene pools of the important crop species. The final crop varieties are not currently generally selected at the centers but sent to collaborators in national programs in many parts of the world, who in turn make the selections that best suit their needs; and many eventually become commercial varieties. Similarly, the centers prepare a series of international crop-yield tests, which include representatives of the best commercial varieties from the world and a few of the most promising experimental lines from collaborators. These are sent to collaborators in thirty-five countries for growing at eighty locations. The data from collaborators are returned to CIMMYT for summarizing and subsequent distribution to scientists in all parts of the world. The data obtained on yield, adaptation, disease and insect resistance in 1 year in such tests are often more meaningful and valuable to scientists engaged in crop research and production programs than data obtained by independent testing at one location for a period of 10 or 15 years.

The international centers also are in a unique position to contribute to practical or internship type of training in all of the scientific disciplines affecting crop production. This type of training is particularly valuable for young scientists from the developing countries because it prepares them for initiating research work upon return to their native country and will also be of value if they subsequently continue their education at the graduate level.

In summarizing the accomplishments of the Green Revolution during the past 3 years, I wish to restate that the increase in cereal production—rice, maize and wheat, and especially in wheat—has been spectacular and highly significant to the welfare of millions of human beings. It is still modest in terms of total needs. Recalling that 50% of the present world population is undernourished and that an even larger percentage, perhaps 65%, is malnourished, no room is left for complacency. It is not enough to prevent the currently bad situation from getting worse as population increases. Our aim must be to produce enough food to eradicate all present hunger while at the same time striving to correct malnutrition.

The Unsolved Problem

To eliminate hunger now in the developing nations, we would need to expand world cereal production by 30%. If it were, however, as simple as increasing the total world production by 30%, regardless of where the production is to be expanded, it could be accomplished rather rapidly by expanding it in the United States, Canada, Australia, Argentina and Russia. But this would not necessarily solve the hunger problem of the developing world, because their weak economies will not permit them to expand their food imports by 30%. Worse still, even if present production could be expanded rapidly by 30% in the developing countries—which I believe is possible based on recent progress of the Green Revolution—so as theoretically to eliminate hunger, the hunger problem as it now exists still would not be solved. There remains the unsolved social-economic problem of finding effective ways to distribute the needed additional food to the vast underprivileged masses who have little or no purchasing power. This is still the great unsolved problem with which the economists, sociologists and political leaders must now come to grips.

I am convinced that if all policy-makers would take sufficient interest in population control and in aggressively employing and exploiting agricultural development as a potent instrument of agrarian prosperity and economic advancement, many of the social ills of the present day could soon become problems of the past. The tropics and subtropics have abundant sunlight and other biological assets and it will be criminal to delay further the conversion of these assets into wealth meaningful to the poor and hungry.

Is It Just?

Some critics have said that the Green Revolution has created more problems than it has solved. This I cannot accept, for I believe it is far better for mankind to be struggling with new problems caused by abundance rather than with the old problem of famine. Certainly, loyalty to the *status quo* in food production—when being pressured by population growth—cannot break the chains that have bound the peasant to poverty and hunger. One must ask: Is it just to criticize the Green Revolution, with its recognized accomplishments, for failure to correct all the social-economic ills of the world that have accumulated from the days of Adam and Eve up to the present? Change we must, or we will perish as a species, just as did the dinosaurs in the late Cretaceous.

The Green Revolution is a change in the right direction, but it has not transformed the world into Utopia. None are more keenly aware of its limitations than those who started it and fought for its success. But there has been solid accomplishment, as I have already shown by concrete examples. I have also tried to indicate the various opportunities for capitalizing more fully on the new materials that were produced and the new methods that were devised. And, above all, I cannot emphasize too strongly the fact that further progress depends on intelligent, integrated and persistent effort by government leaders, statesmen, tradesmen, scientists, educators and communication agencies, including the press, radio and television.

New Cereal Species

But progress is continuous, and we can and must make continuous progress. Better varieties of wheat and other cereals with not only higher yield potential but also with higher content of protein are already in the process of creation.

We need also to explore more fully the feasibility of producing new man-made cereal species with greater production potential and better nutritional quality than those now in existence. Triticale, a man-made species derived from a cross between wheat and rye, now shows promise of becoming such a crop.

During the past 6 years the International Corn and Wheat Center in Mexico, co-operating with the University of Manitoba, has developed a large breeding program to improve Triticale. Within the past 3 years we have developed highly fertile lines and the results up to the present indicate the possibility of combining the desirable characteristics now present in different lines into a single line, thereby creating a new kind of cereal that is superior to wheat in productivity and nutritional quality.

The rapid progress achieved in Triticale improvement suggests the desirability of initiating basic studies to determine the feasibility of developing other cereal species from wide crosses between different existing species or their wild relatives. Recent improvements in individual cell, tissue and embryo-culture techniques, in the development of culture media with additions of hormones and nutrients that foster cell and tissue differentiations, in achieving hybridization between somatic cells and in the methods of inducing polyploidy and mutations, offer many fascinating possibilities of achieving crosses between species that were formerly uncrossable. Even the possibility of using protoplasmic and cell hybridization, followed by manipulation to promote cell differentiation for plant improvement, appears to be nearer.

I propose, therefore, that a bold program of wide crosses be initiated to improve both cereals and legumes (pulses). It should include attempts to make numerous intergeneric crosses among cereals, employing all of the modern techniques to consummate fertilization, and propagate the hybrids. If a series of new combinations can be made and doubled, such as, for example, between maize and sorghum, wheat and barley, or wheat and rice, it would open the door to the possibilities for vast subsequent improvement by conventional methods.

Unfortunately, all cereals are deficient in one or more of the essential amino acids, especially lysine, which is essential for normal body growth, and for the maintenance of health. Protein malnutrition is widespread, especially among children, and many of its victims die or are maimed both physically and mentally for life.

The Ideal Solution

Although food supplements can alleviate this situation, the development of high-yielding varieties of cereal grains that have high levels of protein and better amino acid balance would be the ideal solution, since this would not involve added expense or special educational efforts, and there are good possibilities of producing them. The now famous opaque 2 gene in maize doubles the production of the amino acid lysine which is essential to growth and health in man and many other animals. Similarly, an Ethiopian strain of barley, and some lines of Triticale, have genes for extraordinary production of essential nutrient materials. Plant breeders are trying to combine such genes with the best genes now available for productivity and other desirable characters, thus increasing not only the tonnage of food, but also its essential nutrient quality.

As we are now striving to emancipate ourselves from dependence on artificial food supplements, I have a dream that we can likewise emancipate ourselves to some extent from our dependence on artificial nutrients for the cereal plants themselves, thus lightening the financial burden that now oppresses the small farmer and handicaps his efforts to participate fully in the new technologies.

In my dream I see green vigorous, high-yielding fields of wheat, rice, maize, sorghums and millets, which are obtaining, free of expense, 100 kilograms of nitrogen per hectare from nodule-forming, nitrogen-fixing bacteria. These mutant strains of "Rhizobium cerealis" were developed in 1990 by a massive mutation breeding program with strains of "Rhizobium" obtained from roots of legumes and other nodule-bearing plants. This scientific discovery has revolutionized agricultural production for the hundreds of millions of humble farmers throughout the world; for they now receive much of the needed fertilizer for their crops directly from these little wondrous microbes that are taking nitrogen from the air and fixing it without cost in the roots of cereals, from which it is transformed into grain.

Then I wake up and become disillusioned to find that mutation genetics programs are still engaged mostly in such minutiae as putting beards on wheat plants and taking off the hairs.

Greater Investments

If we are to capitalize fully on the past biological accomplishments and realize the prospective accomplishments, as exemplified in my dream, there must be far greater investments in research and education in the future than in the past.

Few investments, if any, can match the economic and social returns from the wheat research in Mexico. The investment from 1943 to 1964 was estimated to have yielded an annual return of 750%. This study was made prior to the full impact of dwarf wheats on the national production. If the benefits were calculated now, with the inclusion of the returns from the increased wheat production in Pakistan, India, and other Asian and African countries, they would be fantastically high.

Nevertheless, vast sums are now being spent in all countries, developed and developing, on armaments and new nuclear and other lethal weapons, while pitifully small sums are being spent on agricultural research and education designed to sustain and humanize life rather than to degrade and destroy it.

The Green Revolution has won a temporary success in man's war against hunger and deprivation; it has given man a breathing space. If fully implemented, the Revolution can provide food for sustenance during the next three decades. But the frightening power of human reproduction must also be curbed; otherwise the success of the Green Revolution will be ephemeral only.

Population Monster

Most people still fail to comprehend the magnitude and menace of the "Population Monster". In the beginning there were but two, Adam and Eve; when they appeared on this earth is still questionable. By the time of Christ world population had probably reached 250 million. But between then and now population has grown to 3.5 billion. Growth has been especially fast since the advent of modern medicine. If it continues to increase at the estimated present rate of 2% a year the world population will reach 6.5 billion by the year 2000. Currently, with each second, or tick of the clock, about 2.2 additional people are added to the world population. The rhythm of increase will accelerate to 2.7, 3.3 and 4.0 for each tick of the clock by 1980, 1990 and 2000, respectively, unless man becomes more realistic and preoccupied about this impending doom. The tick-tock of the clock will continually grow louder and more and more menacing each decade. Where will it all end?

Malthus signaled the danger a century and a half ago. But he emphasized principally the danger that population would increase faster than food supplies. In his time he could not foresee the tremendous increase in man's food-production potential. Nor could he have foreseen the disturbing and destructive physical and mental consequences of the grotesque concentration of human beings into the poisoned and clangorous environment of pathologically hypertrophied megapoles. Can human beings endure the strain? Abnormal stresses and strains tend to accentuate man's animal instincts and provoke irrational and socially disruptive behavior among the less stable individuals in the maddening crowd.

We must recognize the fact that adequate food is only the first requisite for life. For a decent and humane life we must also provide an opportunity for good education, remunerative employment, comfortable housing, good clothing, and effective and compassionate medical care. Unless we can do this, man may degenerate sooner from environmental diseases than from hunger.

And yet, I am optimistic for the future of mankind; for in all biological populations there are innate devices to adjust population growth to the carrying capacity of the environment. Undoubtedly, some such device exists in man, presumably *Homo sapiens*, but so far it has not asserted itself to bring into balance population growth and the carrying capacity of the environment on a worldwide scale. It would be disastrous for the species to continue to increase our human numbers madly until such innate devices take over. It is a test of the validity of "sapiens" as a species epithet.

An Award for Mankind

Since man is potentially a rational being, however, I am confident that within the next two decades he will recognize the self-destructive course he steers along the road of irresponsible population growth and will adjust the growth rate to levels that will permit a decent standard of living for all mankind. If man is wise enough to make this decision and if all nations abandon their idolatry of Ares, Mars and Thor, then mankind itself should be the recipient of a Nobel Peace Prize, which is "to be awarded to the person who has done most to promote brotherhood among the nations".

Then, by developing and applying the scientific and technological skills of the twentieth century for "the well-being of mankind throughout the world", he may still see Isaiah's prophecies come true: "... And the desert shall rejoice, and blossom as the rose.... And the parched ground shall become a pool, and the thirsty land springs of water..." (Isaiah 35:1,7).

And may these words come true.

The Green Revolution

VIKRAM SARABHAI*

I WOULD like to give an account of what the Green Revolution means to India. But before I do so, and give you some figures and facts, it is perhaps well to understand the local situation under which this is occurring. As everyone knows, India's population now exceeds 550 million. The primary occupation and source of income for a major portion of the population is, of course, agriculture. Unirrigated farms constitute nearly 80% of the total crop area of 138 million hectares. Most parts of India receive rain during the monsoon season, which extends for about three months, from July to September. Of the 138 million hectares which are under agriculture, only about 41 million hectares have an annual rainfall of 1100 mm and above, and about 49 million hectares receive a rainfall ranging from 750 to 1100 mm. Only 20% of all farms resort to multiple cropping.

With growth of population on the one hand and stagnation of agricultural production on the other, the situation had become so serious during the late 1950s and early 1960s that India had to resort to very large-scale importation of food, mostly with foreign assistance. But a remarkable change occurred in the mid-1960s. During the 1951–65 period, wheat from Mexico and rice from Taiwan as well as the Philippines were imported into India for breeding experiments, to see whether these new varieties would take in the Indian soil and produce the remarkable results they had already demonstrated. Side by side with the importation of these new strains from Mexico, Taiwan and the Philippines, Indian agricultural research institutes embarked on an intensive plan to develop new strains for maize, wheat and rice. By 1965 these institutions and developmental agencies were ready to embark on a programme for wide propagation of the new varieties.

One of the first things that was noticed was that the type of grain and the colour of the wheat which had been brought from Mexico, and the quality of the grain of the new rice, were not similar to what the local population had been used to, and the preparations that were produced with many of these were automatically identified as inferior in quality. Hence, even though the productivity of these new imported varieties was good, the lack of acceptability by the population created a problem for their widespread adoption. Fortunately, the agricultural research institutes in

* Former Head, Indian Atomic Energy Project at Bombay, Director of Physical Laboratories at Ahmedabad, India. This paper is based on a talk prepared for the 21st Pugwash conference at Sinaia (Rumania), September 1971. Mr. Sarabhai died in 1971.

India succeeded during the period 1965–7 in genetically building into the imported varieties the desirable characteristics which the local population required. At the end of 1967, a wide variety of improved varieties, which conformed in flavour, colour, and taste to the ones that the people were used to, were in fact available. At that point of time, about 1966–7, the lid was therefore taken off the rapid dissemination of the new technology in agriculture.

The first result of the application of these new varieties (and most of these had to be raised in the irrigated areas) was that yields were doubled; and as soon as yields were doubled and the short-ripening varieties were available, multiple cropping became very much more extensive than it had been before. Production of food grains rose from a trend value of around 85 million tons to 108 million tonnes in a period of about 5 years. And the result of this has been so dramatic that, this year in fact, instead of having a net deficit of food, there is a net surplus. There is already a new problem that people had not fully visualized, the problem of storage of food grains.

The current 5-year plan, which is now at the half-way point, provides for an increase of grain production from 98 million tonnes to 129 million tonnes. Of the planned increase of 31 million tonnes, at least 21 million tonnes are likely to be produced by the use of the new varieties. With the existing pressure on land, the big increases that are taking place are in fact due to these new varieties. This is having, of course, a tremendous impact on employment in the countryside. Of the rural households in India, 52% are small holders and about 24% are agricultural labourers. In the areas where the high-yielding varieties have been sown, the employment has almost doubled. In fact, it is estimated that today, at the end of about 4 or 5 years of intensive work on these new varieties, something like 325 million labour days of additional employment have been created.

What about incomes? One of the most important aspects of the Green Revolution has been that it has reduced the relative disparities between large and small landholders in irrigated areas. In the past, only large holdings used to be able to fetch a viable farm income. With multiple cropping and with these new varieties, even the small holders are now having a profitable agricultural income. I might quote an example from my personal experience: I have a domestic servant who, every morning when I get up at about 5.30 or 6 o'clock and go to my study, brings me tea. He came the other day and said: "You know, I have been working for you for the last twelve years. I don't really know why I am working for you." I said, "Well what's the matter?" He said that he had a small farm of about 4 acres and, about 3 years ago, through a loan he had taken from me, he had made a well, installed a diesel engine, and was now using the new varieties of seeds. He said that the income from his farm was several times greater than the income that he got as a salary earner in my home. What is more, he said that in his community it was improper for his wife, who was living there, to do the farmwork by herself. So he had to engage a neighbouring farmer to do this. That farmer's remuneration was such that it was taking all the cream out of his own farm holding. I said: "Then why do you serve me?" He said: "It is only because every time I am really in need, you give me a loan." This illuminates very well what the Green Revolution is doing to the people.

The disparity, therefore, in the farm earnings of large and small holdings has definitely gone down, very much as a result of the new varieties. But another serious problem is that this has increased the disparity of the earnings of the farmers in relation to the agricultural labourers because, while for both the earnings have gone up significantly, the income on the small farms has gone up even more. So the disparity between the haves, even those with a small plot of land, and the have-nots who only provide manual labour, has indeed increased.

There are some rather important implications of this. Wherever there are large farms, the need for labour during certain seasons is, in fact, so great that mechanization is taking place extensively, as in Punjab and other places. This is regarded as, and indeed in many cases does, running counter to the interests of the farm labourers. Mechanization is still not very widespread, but this is one of the things which is likely to take place and is going to create its own problems.

The increase in real wages has been at least 5% per year during the last 5 years in these areas. This is a very significant aspect of the Green Revolution. But since 80% of the farms are not irrigated, one of the first impacts has been increased disparity in the standard of living and culture between the minority in fortunate areas, which have irrigation, and the large majority of others without irrigation. If, as economists have classically done, we ask the question: Where would fresh investment give us the best incremental return? we nearly always come up with the answer which identifies farmers and communities in the irrigated areas. Investment of governmental funds on this basis would create tremendous imbalances and political instability in the country, as it has already tended to do, through creating some regions far more prosperous than others. To avoid it, one has consciously to adopt a policy of not looking merely for investments that produce greater short-term returns, but consider package investments in new areas, which would then be able to catch up and emerge from their underprivileged position.

A rather interesting study was made recently, by a group of scholars from the Indian Institute of Management at Ahmedabad, on the resistance to change, which is feared might impede agricultural progress. They compared, in a large area in the Western Gangetic Plain, the time at which new fertilizers were adopted and the time when new seeds were adopted. They discovered through this study that, while this gap used to be as much as 4 years before 1965–6, today the gap between adoption of the two different inputs, both of which are necessary for optimum results, is no more than a year. From this they conclude that, at least today, the Indian farmer has no resistance to adopting new agricultural technology, that the return to investment ratio of these innovations is so great all over the country that it has carried its own message with it. There is a fantastic demand for water, for energy to pump water, for fertilizers, and for the new seeds. A most interesting report in the last 2 months is that in several areas traditional varieties of seeds were sold to gullible farmers as being of the new hybrid variety and the fraud is only discovered some time after sowing is completed. The problem of selling genuine new varieties of seeds, which are moreover not infected, is a formidable one. Special measures have been implemented to ensure this.

Dr. Swaminathan, Director of the Indian Agricultural Research Institute, has written quite a bit on the synergistic package that is required before change can be produced in the countryside. In analysing the implications of the Green Revolution, the noted economist Dr. Raj wrote as follows: "New varieties yielded more, only with substantial complementary inputs of chemical fertilizers and other plant nutrients; and they, in turn, required assured supplies of water at specific intervals. Without such supplies, there was danger, not only of the complementary inputs becoming infructuous, but in some cases of the yield being lower than from the use of the older varieties without applying these inputs."

One of the important lessons, that has been learned in the implementation of the measures that have contributed to the Green Revolution is the recognition of the totality of the package that is required. It does not require much imagination to see that if all these inputs are to be brought simultaneously to a particular area, it requires rather new forms of central and local governmental organizations and administrative practices. This is at least as difficult as the task of creating the physical inputs that are required. In the Indian Atomic Energy Commission, we are studying intensively the need for water in large areas where surface irrigation is not possible or is already fully utilized. Fortunately for us, the Gangetic Plain is literally sitting on a lake of water. If one can pump this water out, with deep tube wells or shallow wells, one immediately produces a transformation in the agricultural scene.

It so happens that, in India, fuel resources are concentrated in particular areas of the country; they are by no means evenly spread about. There is at least a third of the country where, even today, atomic energy is less expensive than electric power produced from fossil fuels. Hydro-electricity is, of course, restricted to a few specific locations. Under these conditions, the Atomic Energy Commission has made a study of the type of project that could be implemented if, in an area where there is subsoil water, we could provide extremely low-cost energy. We have completed a study of an Agro-Industrial Complex in Western Uttar Pradesh. The outlines of this are as follows. It covers an area inhabited by 24 million people. It is proposed to put up two nuclear power stations of 600 megawatts each and to energize 25,000 tube wells. The power requirements of tube wells are peaked at different times of the day and at different seasons, and it is proposed to provide a base load for the station, and also to use the surplus capacity of the station, by producing the other inputs required for agriculture: firstly, phosphatic fertilizers, by using the electrothermic process for elemental phosphorus. At the present time, India does not have important deposits of sulphur so that, for producing phosphatic fertilizers using sulphuric acid, all the sulphur has to be imported. Low-cost electrical energy provides a different way of producing elemental phosphorus and this is, itself, not only a better process under the circumstances which we have stated, but is also economical for transportation. Of course, if we have very low-cost energy, then producing nitrogenous fertilizers from electrolytic dissociation of water is an alternative to the use of naphtha or fuel oil, which India has again to import as of now. There is still the problem of making electrolytic cells economical compared to some of the naphtha-based plants. But the growth of technology of high-capacity electrolytic cells has been promising. In consequence, the electrolytic process is likely to at least break even and save foreign exchange. The return on investment in this complex will be 54% on the incremental investment of about 1.5 billion dollars, but 50% of this large investment is on the infrastructures of roads, railways and provision of credit. The complex will generate fresh employment for 1.4 million persons all year round. It will produce 9.4 million tons of food grain, 1.8 million tons of potatoes, 11.2 million tons of sugarcane. It will increase the GNP by 1.2 billion dollars annually. This is the type of complex, in which we have a totality of inputs, being planned to produce the type of change that we desire.

I would like, in this connection, to give an analysis of the formidable social problems that are involved in implementing a project of this type. Such a complex requires effort at a national level, since the setting up of atomic power stations is the responsibility of the Central Government. Moreover, a large station of thousand megawatts capacity for generating electricity would have to be part of a super-grid connecting the important generating points feeding several States. Each State would have to have its own electricity plan, to receive and distribute power to the points at which tube wells would need to be constructed and energized. There would need to be local organizations for digging the tube wells on a co-operative basis and for maintaining them in good working condition. There would be need for hardware, the cables, the distribution towers, switch gear, pumping sets, piping for the tube wells. Unless everything is included in this total framework—from the mining of uranium to the production of nuclear fuel elements, the fabrication of power plants, erection and commissioning of the fertilizer plants, and the whole irrigation system—one cannot hope to achieve the results that the total system is capable of providing. It involves not only the diverse ministries at the centre, but diverse departments of State governments, local governments, industries and co-operatives, not to speak of the effort that would be required from countless private individuals.

We started with the Green Revolution; we are ending up with a social revolution of the first magnitude. In a country like India, provided political power is sensitive to the needs of the underprivileged, provided it takes specific measures to see that the benefits of change do not go into the hands of the few, advanced technology in agriculture, as in power generation, is indeed going to generate a social revolution in our own time.

Knowledge, Agriculture and Welfare

THEODORE W. SCHULTZ*

RECENT large gains in agricultural production in several developing countries have captured the imagination. The additional wheat and rice is a worthy achievement, which has been personalized by a Nobel Peace Prize. The soft phrase, "The Green Revolution", is suddenly popular even among hard-headed scientists. But similar large gains in agricultural production in particular countries (for example, in Mexico) have been achieved since the 1940s.[1] What is new and important is the more general acceptance of the fact that agriculture can be dynamic and progressive in countries where traditional agriculture has been a way of life for many generations. Thus the big chance that has occurred, which has the attributes of a minor revolution, is in seeing and believing that agricultural production possibilities can be transformed from niggardliness to abundance. But there is still some confusion, although it is true that the general public, officials of government and also scientists are now seeing with increasing clarity that advances in scientific knowledge can contribute to the process of increasing agricultural production, and thereby enhance the welfare of people.

The main reasons for the residual confusion are the following: (1) An over-simplification of the linkage between the Green Revolution and the sciences. (2) A lack of understanding of necessary economic requirements for the efficient adoption and use of the new, superior agricultural inputs. (3) Seeing the welfare implications of such gains in agricultural production in terms of the detailed effects upon different classes of farm people in an area, and not seeing the more general effects of cheaper food grains on the welfare of a total population. (4) A lack of clarity about ways of developing efficient systems of agricultural research, in spite of many decades of

* Professor of Economics, University of Chicago. Based on a paper presented at the 21st Pugwash Conference at Sinaia, Rumania, in September 1971. It also appeared in *Science Studies*, **2,** 361–8 (1972).

[1] In Mexico, between 1946 and 1962, the yield per hectare of the harvested area devoted to 37 crops doubled inside of the irrigated districts and it increased 40% outside of the irrigation districts. The production indices of key crops show the following increases with 1960 = 100:

	Wheat	Rice	Corn	Cotton
1946	31	58	63	23
1962	110	129	107	118

Although the rate of population growth has been high, exceeding 3% per year, the per capita consumption of farm products in Mexico increased by 43% between 1940 and 1965. From Reed Hertford's "Sources of Change in Mexican Agricultural Production 1940–65", unpublished Ph.D. dissertation, University of Chicago, March 1970.

experience with organized agricultural research, and the worthwhile lessons that could be drawn from this rich experience.

I shall elaborate briefly on each of these four reasons for the confusion that still continues to hamper and delay the modernization of agriculture.

Dependence of the Green Revolution on the Sciences

The Green Revolution reveals clearly the value of organized agricultural research. It demonstrates once again that nothing succeeds like success. But what is not revealed is that the new varieties of wheat and rice are only a part of the contributions of the sciences. Modern agriculture would be impossible were it not for advances in scientific knowledge. From an economic point of view, this knowledge, when it has become transformed into inputs, techniques and into the skills of man, is the most important factor in production.

There is a long history of pre-science agriculture, during which there were some innovations, some discoveries and some learning from trial and error. But pre-science agriculture, then and now (for it still prevails in many parts of the world), is at the mercy of the vicissitudes and niggardliness of nature. Raw land and brute labor are not sufficient to dispel the ancient fear of starvation, out of which arose the age-old prayer, "Give us this day our daily bread". Surpluses of agricultural products are a recent development; Mexico now has a surplus of wheat, and even Japan has a surplus of rice. Thus the niggardliness of nature is transformed into abundance, and the sciences are playing an important role in that transformation.

The first lesson of the Green Revolution is that modernization of agriculture is dependent upon a wide array of scientific knowledge—physics, chemistry, and biology; and also upon engineering, with respect to irrigation and transport facilities, and agricultural machinery. There is, however, a strong tendency, which is all very human, to claim too much for that bit of new knowledge about which the particular scientist or engineer is an expert. The second lesson, which is not obvious, is that the large gains to be had from the improvements in the technical production possibilities of agriculture depend in large measure upon the combinations of technical factors and the gains from the *interactions among them*.

Even in undergraduate chemistry, one learns about interactions. It is obviously not a new idea in the sciences. R. A. Fisher, distinguished for his small-sample theory, designed agricultural field experiments to control for, and to take account of, interactions. Charles Kellogg, the soil scientist, extended the concept to the more aggregative interplay between the state of the soil, a new variety, additional fertilizer, better control of water and more effective pest control. Several inferences can be drawn from these interactions.

(1) Norman Borlaug would be the first to agree that the new dwarf wheat varieties developed by his Rockefeller–Mexican enterprise require fertilizer, water management, control of pests and a lot of dry-season sunshine. Similarly, in attaining these gains in agricultural productivity, fertilizer is not a panacea, no matter how

cheap it may become or how abundant the supply. Thus, neither the highly productive new varieties, nor the cheap fertilizers, are by themselves sufficient, although the contributions of the plant breeders and of the engineers in developing the high compressor processes to produce cheap nitrogenous fertilizer have both been necessary parts of the process. So have been the contributions of irrigation engineers, although these engineers tend to have a built-in bias for large systems and therefore neglect small projects and tube wells. When it comes to pesticides (which are also necessary) the types of pest control that will be required in conjunction with the new agricultural inputs in India and other developing countries are as yet, in large part, not known.

(2) The popular urban view, which includes the view of scientists who think of agricultural scientists as second cousins, is that modern agriculture is necessarily soil depleting. But in general this is not so. The "interaction" between farmers and soils is usually beneficial; the goodness of the earth as agricultural soil is in substantial part man-made. For example, the renowned soils of Iowa are presently better than they were, say, four decades ago; average corn yields since then have risen from around 30 to 100 bushels an acre. But this "interaction" also precedes modern scientific farming; even in India, after centuries of cultivation, the soils are, in general, presently better than they were a century ago. Moreover, as a consequence of the Green Revolution the productivity of the soil in the Punjab is becoming better still. The crisis-bent ecologists would do well to ponder this important fact.

(3) A closely related interaction concerns the effective supply of land. It is not true, as all of us were taught, that the supply of agricultural land is fixed once and for all by the surface of the earth. On the contrary, it is a fairly elastic economic variable over time, depending on the "interaction" between new knowledge, additions of complementary forms of capital and the value of farm products. It can be augmented in many different ways, but mainly by increasing its capacity—in much the same way as the aeroplane now carries several times as many passengers as it did in earlier years. Thus, for example in the United States, the increases in crop production per acre have been so large that over 50 million acres have been taken out of crops during the last several decades. A similar development is under way in parts of western Europe. When the agriculture of India has become fully modernized, India also will require less crop land than the acreage presently under cultivation, provided the serious high rate of population growth is checked.

In short, the old maxim of turning swords into ploughshares should now be interpreted to mean turning defence-oriented science towards agricultural scientific endeavors.

The Economic Part of the Story

Turning next to the motivation and capability of farm people in adopting and in using efficiently the new scientific inputs, there are several necessary economic requirements. These requirements include efficient economic incentives, adequate information, and the learning of new farming skills. Throughout much of the world, in both rich and poor countries, farm people are thwarted by distortions in economic incentives and by a lack of information. Moreover, the process of modernization is delayed because of the time that is required to learn the new farming skills (mainly due to the low level of schooling).

It is unfortunately true that the dominant élite view, both in capitalistic and socialistic countries, tends to be that farmers are dullards, inherently backward, strongly inclined to farm in accordance with tradition, and fundamentally indifferent to any new and better economic opportunities. Thus, they have long been maligned. One of the main purposes of my book, *Transforming Traditional Agriculture*,[2] was to lay to rest this widely held, mistaken view about farm people. The Green Revolution has undoubtedly accomplished much more than my book did in establishing the fact that farmers in Mexico, or in India, or in any other country, are not indifferent to real opportunities to improve their economic lot.

But the trouble is that governments are slow to learn. Agricultural policies once established are then maintained by the vested interest of the bureaucracy that administers the farm program. Land reform is as a rule promoted to attain particular social and welfare objectives. But land reform without the scientific inputs is not sufficient to attain large gains in agricultural production. Land reform, even with scientific inputs available to farmers, but without economic incentives, is also unsuccessful. Similarly, the reorganization of agriculture into large farms, along with modern tractors and machinery, fertilizer, new varieties of seeds and pest control, but without effective economic incentives that reward farm people for adopting and using these new inputs efficiently, is somewhat less than successful. The governments of most of the rich capitalistic countries have made the mistake of establishing inordinately high price supports for farm products that require vast subsidies. The scientific inputs have been available; the economic incentives to apply them have been very strong. But the economic distortions caused by the price supports and subsidies are serious and very costly in terms of unwarranted "surpluses", of high taxes, high cost of food to consumer, and of the regressive effects that this policy has had on the distribution of personal income within agriculture. Thus this policy, too, is not successful in providing the essential economic requirements.

The gains in agricultural production associated with the Green Revolution also have several important economic implications that should not be overlooked:

(1) The increasing dependence on imported food grains of some of the major densely settled poor countries that occurred during the early post-World War II period, is being reduced markedly by the gains in their agricultural production. The OECD view, expressed as late as the mid-1960s, is no longer tenable. A few of these less-developed countries are now prepared to export some farm products to the European Economic Community countries, but the barriers that the EEC has erected to keep such imports out makes it very difficult.

[2] New Haven: Yale University Press, 1964.

(2) The increases in production of food grains give the less-developed countries some additional, much-needed time in which to cope with the complex problems of excessively high human fertility and population growth.

(3) The entire economy of these countries is much more dependent upon the gains in agricultural production than is the case in the rich countries. The possibilities of attaining general economic growth in the less-developed countries is measurably enhanced by these agricultural gains.

(4) Although in many less-developed countries the ratio of farm workers to the area of land under cultivation is still rising, it would be an error to infer from this that the modernization of agriculture cannot proceed. The scientific agricultural inputs are, in general, highly divisible: a bushel of high-yielding dwarf wheat, a bag of fertilizer, and a few pounds of pesticides. Similarly, small farm machinery and garden tractors can be had, and can be efficient, as the farmers of Japan and Taiwan have demonstrated.

(5) Last, but not least, is the important implication that the agricultural productivity from the advances in knowledge means cheaper food, and a rise in the real income of consumers. In a well-managed economy, these gains become a consumer surplus.

The Green Revolution and Welfare

Much of what is being written on the welfare implications of the Green Revolution is wrong. There are two unsettled questions. Who is to be the judge in determining welfare? What is the evidence? The thrust of the misleading welfare argument against the Green Revolution runs as follows: although the gains in agricultural production are being attained efficiently on narrow economic grounds, they are bad in terms of welfare. It is asserted that rich farmers profit and poor farmers become poorer; landless laborers become even worse off than they were before; and the welfare of the country is impaired.

Welfare considerations are indeed important. There are, however, two pitfalls in the path of the outside observer, given his standard of values and ideology. The welfare objective of, for example, the people of India, is an internal matter. It cannot be determined by the values of outsiders and their preferences. It is up to the people of India to decide what their welfare objective should be. The other pitfall is in the inaccuracy of the evidence that is being presented on the economic consequences of the Green Revolution. In the Indian part of the Punjab, where the new highly-productive wheat varieties are doing exceedingly well, it is true that those with "huge" farms (i.e. of up to 20 acres!) are in clover. But what is overlooked is that the farmers with 5 acres, or 3 acres, or even less, are also benefiting from producing more wheat. The scattered evidence now available suggests that substantially more labor is now required in the Punjab, and, in real terms, farm wages have risen appreciably. It is undoubtedly true that given time the in-migration of farm workers from other parts of India will beat down these wages to the very low levels that characterize farm wages generally in that country. But the most important consequence of the Green Revolution is lost sight of—that is, the *consumer surplus* that it is producing for India. Cheaper wheat means that the real income of consumers rises. Farm laborers also benefit from this consumer surplus. Thus the evidence pertaining to the welfare effects of the Green Revolution is not what it is presumed to be by those who are belaboring the particular, local welfare consequences of these gains in agricultural productivity. A major issue is, however, emerging.

The opportunity to modernize agriculture is as a rule quite unequally distributed within any large country. The heterogeneity of the agricultural production possibilities, in a large country, sets the stage for income disparities between areas, and for differences in population pressures among farming regions. What this means is that the modernization of agriculture alters significantly the comparative advantage among agricultural areas within the country. Western countries are not exempt on this score. Clearly, parts of agriculture in Italy and France have long been depressed. The U.S.S.R., despite her centralized planning and administered economy, is not spared, and the depressed Appalachia is poignant testimony of the very uneven agricultural development in the United States. Among the poor countries, Mexico and India illustrate the implications of this development. Mexico, which is now well into the third decade of successful modernization of parts of her agriculture, faces increasingly serious income disparity between major agricultural areas. Agriculture in north and middle Mexico has progressed, but the less successful south of Mexico has, by comparison, become a major depressed area. Internal migrations have profound implications in terms of the redistribution of the Mexican population. In India, the agricultural comparative advantage of the producing areas is shifting to the northern parts, and to the major "rice bowls" of the south, as a consequence of modernization. A large triangle in Central India is losing out competitively. Scores of millions of people who are dependent upon agriculture reside in this large area. They will be left behind. The new varieties of rice, wheat, corn, millet, etc., that are responsive to the larger and cheaper supply of new fertilizers, are decisively less productive within this large triangle because of the lack of rainfall and of water for irrigation. The burden of a lecture that I presented at Delhi University in 1967 was directed to this economic and social problem, which would emerge in India, and my plea was that it should be high on the agenda of Indian economists.

Enlarging Organized Agricultural Research

It is obvious that the Green Revolution would have been impossible without organized agricultural research. The remarkable gains in agricultural production, especially since World War I, mainly in the scientifically advanced countries, could not have taken place without this research. Organized agricultural research has a long rich history. The system of land grant colleges and experimental stations that was established in the United States fully a century ago has been generously supported with public funds. (By comparison, the use of significant public funds to sponsor organized research in other parts of the sciences is a much more recent development.) European countries, Canada, Australia, Japan and notably Russia, have long histories of successful organized agricultural research. But the less-developed countries have been much less fortunate; for they have not, in general, benefited until very recently from this important development. Moreover, they are still benefiting all too little in view of the high pay-off to be had from investment in agricultural research.[3]

With special reference to the less-developed countries, why has there been so little progress since World War II, in view of the efforts of the United Nations, including FAO, and of many governments purusing bilateral approaches? The following reasons are suggested:

(1) In part, it is a consequence of the overwhelming importance that has been given to national defence-oriented scientific research. This has usurped an inordinate amount of the increases in research funds.

(2) While governments in developed countries tend to give agricultural scientists within their own countries substantial authority in selecting research projects, when these same governments provide funds for agricultural research in the less-developed countries they tend to sponsor narrow, inefficient mission-oriented research. Moreover, the support is subject to the vicissitudes of foreign policy; it therefore lacks the long-run stability that is absolutely necessary in developing effective organized agricultural research.

(3) The agencies of the United Nations, including FAO, despite good intentions, have not succeeded in establishing centers of sufficient scale, and with sufficiently long-run horizons, to attract highly competent agricultural scientists. (There are, of course, some exceptions.)

The one notable success story has been the agricultural research enterprises sponsored by private foundations. The most outstanding, in terms of achievements, has been the Mexican–Rockefeller research establishment in Mexico, which was launched in the early 1940s, and to which the Ford Foundation is currently also contributing funds. "Borlaug" and "dwarf wheats" have deservedly become familiar words. The International Rice Research Institute, to the credit of the Philippines and the foundations, is another success story. There are also several others that are doing well, and a major project of this kind is getting under way in Nigeria.

What this brief review adds up to is rather discouraging. Governments of countries

[3] The recently completed analysis by Nicolas Ardito-Barletta of the costs and returns from the Mexican–Rockefeller Foundation agricultural research programme, covering the period from 1943 to 1964, leaves little room for doubt that this programme has been an exceedingly profitable investment. Capital is scarce in Mexico inasmuch as the rates of return to capital range between 15% and 25%. This agricultural research program up to 1964, analyzed as an investment, shows an internal rate of return between 54% and 82%, more than twice as high as the rate of return on alternative investment opportunities. (See Nicolas Ardito-Barletta, "Costs and Social Benefits of Agricultural Research in Mexico", unpublished Ph.D. dissertation, University of Chicago, June 1971.)

that are rich in science do badly in assisting less-developed countries in establishing organized agricultural research. Nor have the national academies of sciences in the rich countries developed realistic and meaningful plans that would maximize the division of research between rich and poor countries. But the best of plans will not suffice if only foundation funds can be had, because the foundations simply do not have sufficient resources for this task. Since the rewards from organized agricultural research, in terms of consumer welfare, are very high, it behoves the governments of the less-developed countries to enlarge greatly their own investment in such research.

My overall conclusion is that the Green Revolution will turn out to be a godsend, provided we learn from it how important agricultural research can be, what the economic requirements are for the adoption and efficient utilization of the results of such research, where to look for the gains in welfare, and what must be done to extend the benefits from organized agricultural research to all of the less-developed countries. Once we learn and act accordingly, the age-old niggardliness of agriculture will be transformed into abundance, provided of course, that human fertility is checked.

*Extending the Green Revolution**

Sterling Wortman†

On August 1–3, 1973, representatives of twenty-nine governments, assistance agencies, foundations and developing regions of the world met in the Board Room of the World Bank in Washington. They were there to try to put together $30 million for 1974 support of international agricultural research. They did marshal the money, up from $15 million in 1972 and $23 million in 1973. They also launched the seventh and eighth in a series of strategically located, new types of research and training centers. These are considered crucial to acceleration of basic food crop production in some eighty food-deficit countries and to improved standards of living of hundreds of millions of poverty-stricken farm families.

This Consultative Group for International Agricultural Research, sponsored by the World Bank, UNDP and FAO, was meeting without fanfare as has been its custom. But its present efforts may be ranked by future historians as among the most important of the century in man's struggle to balance food supplies and burgeoning populations of the poorer countries. Its activities hopefully will buy time needed to stabilize population growth rates and contribute to gains in per capita output of food and increases in standards of living generally.

The events that led to formation of the Consultative Group are unique in agricultural history. Some authorities see in them a major breakthrough in developmental assistance—a pattern of international cooperation which, if successful in agriculture, could be adapted to other problems of public nature such as population and health, education, or environmental protection.

To understand the forces which are bringing together this unlikely combination of governments, international agencies and private organizations, one must look at the nature of the problems they seek to solve.

The Nature of the Problem

At present, two-thirds of the people in the world live in the developing countries. During the 1970s, over 400 million additional individuals will be added to the already burgeoning populations of the nations of Asia—a number approaching that of the combined populations of the U.S. and Russia. Most of these 400 million or

* First published in *World Development*, Volume 1, No. 12, December 1973.

† Vice-President, The Rockefeller Foundation, New York.

so infants will be born to rural families since 70–80% of the people of these nations live and work on farms or in rural trade centers.

While numbers will differ, a similar situation confronts most countries of Africa and Latin America.

Most farm families of developing countries have long been dependent for food and minimal income upon meagre crop yields from long-impoverished soils. Many are subsistence farmers in areas remote from centers of commerce. They neither produce much for urban markets nor purchase products of urban industry. Primarily concerned with their own maintenance, they contribute little to the long, tough task of nation building. Per capita incomes of such rural people commonly are abysmally low—well under the national averages of \$100 to \$200 a year, a small fraction of the \$3900 average of the United States. These rural dwellers suffer from inadequate diets; they lack access to health, education and other social services—services which they cannot afford to purchase and which cannot be financed by governments operating on shoestring budgets.

Farms are for the most part tiny by American standards. And, since relatively small amounts of additional land can be brought into cultivation, farms tend to become fragmented still further as lands are divided among heirs. Given the lack of opportunity in the countrysides, the young flee to the cities.

The subsistence farmers who depend for their livelihood on the basic food crops—cereal grains, root crops, food legumes and several animal species—have been among those with the most desolate of futures. While for plantation crops (rubber, sugarcane, bananas, pineapple, to name a few) governments or commercial companies have supported research, supplied necessary supporting services, and organized markets, there had until the mid-1960s been little such attention given to the improvement of the basic food crops in the developing countries. Yields of rice, wheat, corn, barley, sorghum, millets, field beans, chick-peas, pigeon peas, cowpeas, cassava, yams, sweet potatoes, and animal species remained low and static at levels of decades and centuries past. Meanwhile, mostly out of sight of central governments, the people suffered.

The era of neglect hopefully is near an end. With it may come a period in which the hopelessness of hundreds of millions of rural people will be replaced by well-grounded expectations of a better future.

Behind the sense of hope lie a number of quite recent and highly significant agricultural developments generally missed by the press. They involve new commitments of developing nations, new institutions to serve them, new forms of international co-operation, and a new air of confidence that progress is possible.

The commitment of nations is based on the recent realization that increased productivity of farms leading to increased incomes for rural people is fundamental to total national economic progress, that in many regions rapid intensification of agriculture can be achieved through science.

It is now clear that the technology required for intensive crop or animal farming, particularly the biological components, must be developed in the regions where it is to be used, tailored to conditions on the farms. There is no other recourse. Crop varieties and associated production methodology must be adapted to the peculiar combinations of soil and climate where they are to be grown. Means of control of local insects and diseases are essential. Produce must meet consumer requirements, whatever they may be. Obviously, while scientific methodology is transferable from developed to developing countries, the technology itself is not.

Given the substantial number of important unimproved basic crops, the sophistication of the science involved, the diverse agricultural conditions found in most countries, the weakness of national institutions, and the paucity of trained people, the task of modernizing agriculture facing each nation is both massive and complex. To this one must add that changes must be made by vast numbers of individual farmers, many poorly educated if at all, and most with land holdings of a very few acres.

Multiply this situation by some eighty countries having pressing but different needs. Consider trade, ideological and other rivalries among nations which make co-operation among them difficult. Compound the complexity with language barriers, the ever-changing nature of biological problems, and the fact that the war on hunger and poverty will involve millions of continuing separate battles on little farms. Add in the shortage of funds available through assistance agencies and unreliability of appropriations by governments of affluent nations. The result is a world problem of major proportions.

This was the dilemma faced by leaders of fifteen national, international and private organizations who met informally just a few years ago (April 23–25, 1969) at The Rockefeller Foundation's Bellagio Study and Conference Center in Italy.

A Basis for Action

For 3 days the institutional leaders met informally and off-the-record, listening to evidence concerning the gravity of the world food situation and to proposals for accelerating progress. They discussed strengths and weaknesses of existing assistance efforts, and reached consensus on a number of important points.[1]

They concluded that gains in agricultural production of at least 4% per year are needed for the rest of this century, given present rates of population increase, if even

[1] Participants included David Bell of the Ford Foundation, Adekke Boerma of FAO, Fred Fournier of France's ORSTOM, Robert Gardiner of the Economic Commission for Africa, John Hannah of USAID, George Harrar (the host) and Will Myers of The Rockefeller Foundation, Paul Hoffman of UNDP, Robert McNamara of the World Bank, Edward M. Martin of OECD, Ernst Michanek of Sweden's SIDA, T. Ohuchi of the Asian Development Bank, Masao Sawaki of Japan's Economic Cooperation Bureau, Maurice Strong of the Canadian International Development Agency, Sir Geoffry Wilson of the United Kingdom's Ministry of Overseas Development, and Al Wolf of the Inter-American Development Bank. Consultants included K. L. Bachmann and Jose Vallega of FAO, William Clark of the World Bank, Lowell Hardin and F. F. Hill of the Ford Foundation, Paul-Marc Henry of UNDP, and the author for The Rockefeller Foundation. Six position papers and a statement of conclusions reached were published by The Rockefeller Foundation as *Agricultural Development, Proceedings of a Conference.*

modest improvements in standards of living of the world's masses of poorer people are to be realized. On an annual basis this may not seem difficult, but it involves a doubling of output in 18 years and a quadrupling in 30! One recurring question was, "Where are the men and the institutions to successfully mount such an effort?" It was realized that most of the work would have to be done and financed by the developing nations themselves. But most of the nations lacked trained manpower and had no hope of mounting in time the sophisticated scientific effort needed to modernize agriculture.

Fortunately, there was evidence that nations could, given the will, make rapid progress, particularly with new types of developmental assistance becoming available. Participants noted that in 1943, with technical assistance (eighteen scientists at the high point) from The Rockefeller Foundation, Mexico's Ministry of Agriculture undertook through a newly created "Oficina de Estudios Especiales" to develop the needed technology for Mexico's main food crops, to train the hundreds of technical personnel, and strengthen the governmental and private sector institutions necessary for dynamic agriculture. It was successful. By 1963 Mexico's scientific personnel were fully in charge of their own technical institutions and the then president of The Rockefeller Foundation, Dr. J. George Harrar, was able to state that Mexico could stop importing wheat in 1956, when harvest balanced demand for the first time in history, and corn was no longer in deficit supply.

The next dramatic examples of national progress were those of India and Pakistan. In the mid-1960s, after two successive bad crop years, they were seemingly faced with the bleak prospect of severe food shortages for growing populations and inadequate financial resources to purchase supplies abroad. The Paddock brothers' *Famine 1975* predicted widespread starvation. However, two developments in India and Pakistan indicated they may have been wrong. First was the timely availability of short, stiffstraw, high-yielding wheat varieties developed by the Mexican and Rockefeller Foundation scientists, and of similar high-yielding rice varieties from Taiwan and the newly created (1960) International Rice Research Institute in the Philippines. The second was the clearly demonstrated capability of the governments, contrary to many predictions, to bring rapidly into being new policies, programs and institutions so that the new technologies could reach millions of their farmers, many having small land holdings.

These significant and rapid developments in Mexico, India and Pakistan, and in the Philippines, Turkey and other countries, were popularly captioned the "Green Revolution."

International Centers: Catalysts of Cooperation

There had been one other unique and important development which interested the Bellagio Group. This was the invention of the privately organized but international institute.

By 1959 George Harrar of The Rockefeller Foundation and F. F. Hill of Ford were distressed that little was being done to improve rice for the people of tropical

Asia. Rice was and is the staple food of most of Asia's millions. But prospects were nil that each nation could have a team of research scientists after the Mexico model. There simply were too few capable scientists in Asia and it would be too expensive to supply them from elsewhere. They then struck on the idea of creating *one* scientific team of the world's best and most dedicated scientists, without regard to nationality, that would work for all nations, helping to solve major problems limiting productivity of tropical rice and training personnel for Asian institutions. Authorities of Asian countries were asked for their reactions. Yes, they said, such an institute was needed. No, they said, financial support by any one country was unlikely unless the institute were located there.

To shorten the story, the foundations, after discussions with the government of the Philippines, agreed to create a totally new kind of institution—an international, nonprofit, research and training institution, with an international staff. It would be governed by an apolitical, self-perpetuating board of trustees with a majority of the members being eminent citizens of the nations to be served. The initial staff of eighteen consisted largely of young (25–35) award- or honors-winning scientists. They came from eight different countries. They had up-to-date training and a sense of mission. That mission was to help the nations of Asia get their rice yields up as quickly as possible. The staff generally agreed with the need for excellence of research on relevant problems, an interdisciplinary approach, the training of large numbers of Asian scientists and technicians, the publication of results, and the stimulation of cooperation among Asian institutions. However, they understood and accepted that their success would not be measured by these conventional factors. They knew that unless the rice yields of Asia rose, the International Rice Research Institute (IRRI) would be judged a failure.

The record of IRRI is 12 solid years of success. By 1965 its breeding program had created a sturdy dwarf rice, IR8, which gave such high yields when properly fertilized and watered that observers called it "miracle rice". IRRI breeders objected to so glib a term for a scientific achievement, but the name stuck. IR8 soon had equally prosperous cousins, IR5, IR20 and IR22, and still other varieties followed, bred either by IRRI or by rice breeders in other countries who crossed IRRI rices with their own strains. These new rices carried genes for disease resistance, high yield, stiff stalks, the ability to make the most of fertilizers, and other virtues. Two prime attributes were their insensitivity to changes in length of day and their shortened growing period. This meant that planting times did not have to be synchronized with the seasons, and in the subtropics farmers could grow three crops a year—or two crops of rice plus a crop of beans or corn or yams. The new rices required more care and more investment, but they also paid higher dividends.

The Philippine Government got behind rice-production programs, promoting the use of IRRI varieties and the "packages" of recommended farming practices that went with them. A vigorous campaign resulted in rice sufficiency in 1967 for the first time in 60 years; by 1968 they were exporting rice. Subsequent reverses have darkened this picture, proving once again that agriculture is only one element in a larger configuration and is still very vulnerable, however sophisticated its technology.

In the late 1960s the new IRRI rices spread rapidly throughout South and Southeast Asia, wherever governments were willing to back them with credit and extension programs. A particularly successful effort was the All-India Coordinated Rice Improvement Program. IRRI varieties were also adopted or used in breeding locally adapted rices in Africa and Latin America. Fifteen such rice varieties have been released in other countries to date. IRRI scientists are now concentrating on obtaining even higher yields, on improving nutritional quality of the crop, and on conquering major diseases and pests of rice. Several promising lines have been developed that have resistance to more than one plant enemy, and a newly released variety, IR24, yielded a record 11 tons per hectare in the 1972 dry season—1 ton more than IR8's "miracle" yield.

A training program for rice scientists is one of IRRI's most important features. Every level of training, from country agent to postdoctoral scientist is offered, and IRRI know-how is spread rapidly to other nations through its system of training the trainers. Key men are sent by their governments to study rice production in the Philippines, and then they return to teach IRRI's techniques to their own production managers and extension people. IRRI also sends out staff members on special assignments to help other nations launch breeding and research programs or get production efforts organized.

Two special programs under way at IRRI may prove as crucial to feeding Southeast Asia as the development of superior rices: these are the multiple cropping program, developed by Dr. Richard Bradfield, and a recent effort to raise yields of rice on small farms that have no irrigation. The multiple cropping idea, while not new to tropical-climate farmers, was given a new impetus by the concept of genetic manipulation or "biological engineering", used by IRRI to breed plants tailored to meet specific needs. The new rices were one important element; Dr. Bradfield and his co-workers also developed economical systems in which a second crop was planted between the rows of the one that was ripening, taking advantage of residual moisture or fertilizer, or benefiting from the special traits of the plant itself—for example, the nitrogen-fixing property of legumes would give the next crop a boost. Intercropped corn and sweet potatoes have been a great success.

The upland rice program is using the whole spectrum of IRRI's resources, its agronomic techniques and extension methods in particular, to study ways of raising yields of rice on subsistence farms in drier areas. These relatively unprofitable operations occupy about 70% of all the rice land in the Philippines. The program is still experimental, but to date the results in about 150 widely scattered trial projects are very encouraging.

From its inception, IRRI had one other important feature. It could take funds from any source as grants or for contracts. It could, on invitation, place people anywhere or undertake any activity which its Board thought would be useful and for which it had funding. Soon, it had staff members at work at national institutions—at the request of the governments concerned and with supplementary funding by assistance agencies—in India, Pakistan, Ceylon (Sri Lanka), and Indonesia. It was cooperating closely with institutions in most other Asian countries.

Scientists of Japan, India, the United States, the United Kingdom, the Netherlands, and Australia—wherever the expertise existed—found they could cooperate with IRRI's competent staff at high scientific levels, thereby enabling their findings to be applied to problems of many rice-growing nations.

Thus developed the concept of international networks of scientific effort on problems such as rice production—networks of specialization (or individual authorities) in developed nations, the international institutes, organizations in developing countries and through them their farmers. This would permit expertise wherever located in the world to be applied to problems wherever they might exist, and without undue delay.

In 1966 the Ford and Rockefeller foundations joined forces with the Government of Mexico to create the International Maize and Wheat Improvement Center (CIMMYT from its Spanish name) to do for wheat and corn worldwide what IRRI was doing for rice. CIMMYT did not have to build from the ground up as IRRI did; it was an outgrowth of the well-advanced international programs in corn and wheat improvement that had developed out of The Rockefeller Foundation's co-operative Mexican Agricultural Program. A great deal had already been accomplished in the creation of high-yielding, disease-resistant varieties of wheat and hybrid corn; training programs were also under way. CIMMYT's status as an international institute consolidated this work and expanded its scope.

CIMMYT made world headlines in the late 1960s when Mexican wheats sparked the Green Revolution in India and Pakistan. CIMMYT's role in this effort was greatly facilitated by its international charter, which left it free to work with other organizations, such as the Ford and Rockefeller foundations, USAID, United Nations agencies, and the governments concerned. Similarly, when CIMMYT scientists developed experimental strains of a new high-protein corn, based on a mutant gene called opaque-2, the United Nations Development Programme became interested and provided CIMMYT over $4.0 million to intensify the work and to extend it to other developing areas. Opaque-2 corn and another mutant called floury-2 have relatively high percentages of lysine, an important protein constituent in which ordinary corn is deficient. This characteristic excites nutritionists as well as public health experts and national leaders in developing countries, because no one has yet solved the problem of getting enough protein into the diets of poor rural people who live mainly on corn and as a direct consequence suffer from malnutrition, with all its attendant ills. Corn breeders may have the solution in high-lysine corn. CIMMYT scientists recently reported that the release of a competitive commercial variety is likely to be announced in the near future.

Another CIMMYT project that will make a real difference to the poor man is a 6-year-old effort to raise yields on subsistence farms in a corn-raising district in the State of Puebla in Mexico. Using farming techniques worked out by CIMMYT and Mexican agronomists, and extension and communications techniques developed by teams of specialists working closely with local leaders and farmers, the Puebla Project succeeded in doubling yields on 3000 farms by 1970; 8441 farmers are taking part in the Project this year, up from 103 in 1968. When corn harvests begin to bring in enough cash income, Project leaders plan to teach the farmers to raise other crops; as they gain the confidence of the people, they hope to generate community interest in health and sanitation, farm economics and nutrition, and other means of improving the quality of rural life.

The UNDP and the Inter-American Development Bank are providing funds to train extension workers from other parts of Latin America, who plan to set up similar projects; already the State of Mexico has instituted an ambitious program based on the Puebla pattern. In the light of estimates that between 55 and 90% of the farms in different regions of the developing countries are small subsistence-level operations like those in Puebla, the crucial importance of such efforts is clear. The fact that the demonstration was undertaken by an international institute, primarily for the purpose of developing a strategy, gives it the necessary leverage for application elsewhere. International conferences have been held to acquaint leaders from other countries with the Project's principles; studies of its findings have been published and circulated; data have been carefully recorded for future economic and social analyses as well as further technological research.

The international-institute concept was carried one step further in 1967 when two new centers were founded. CIAT, the International Center of Tropical Agriculture, was established in Colombia, with support from the Ford, Rockefeller, and W. K. Kellogg foundations, and the International Institute of Tropical Agriculture (IITA) was founded in Nigeria by the Ford and Rockefeller foundations. CIAT is the only institute so far to have a well-developed animal sciences program, with research on beef cattle and swine for tropical environments. It is also concentrating on improvement of manioc or cassava, a major food crop throughout the lowland tropics of Latin America and the Caribbean region, and on grain legumes and soybeans. IITA is focusing research on developing farming systems for tropical Africa and on improving such crops as cowpeas, yams, and sweet potatoes; it is also promoting local and regional production of rice, maize, pigeon peas, soybeans, and cassava.

Looking back on these developments, the Bellagio Group concluded at that first meeting in 1969 (since dubbed Bellagio I) that the development of technology for the developing regions of the world, the strengthening of national institutions, and the creation of international communities of effort on major problems probably could be achieved most rapidly and economically by extending and deepening the programs of the existing four international centers and by creation of any additional ones needed.

The Group recognized, however, that the major effort would have to be made in and by the developing countries; only they can set the policies, organize the institutions, establish the programs necessary to reach their farmers. The institutes receive more requests from governments for help with agricultural projects than they can reasonably fulfill; their training programs and the provision of consultants are the mechanisms through which they give assistance where they cannot conduct programs themselves. International scientific workshops, seminars, and conferences, plus an active network for exchange of information and sharing of breeding materials, also strongly backstop programs in cooperating nations. For example, CIAT is extending aid to Guatemala and to Ecuador through a system of consultants and trainees; CIMMYT is training wheat specialists for Latin American, North African, Asian and Near and Middle Eastern countries; IRRI has a world-world alumni network of rice scientists leading production programs in farflung places. The attitude of most national leaders toward the institutes was expressed recently by the Governor of the Bank of Venezuela who spoke for all the countries of Latin America and the Philippines at the annual meeting of the World Bank: "The action taken in establishing the Consultative Group on International Agricultural Research, with the Bank sponsorship and participation, has already received our support and applause. We should like to take this opportunity of reiterating our approval because we believe that basic research in this sector is of fundamental importance for development; we trust that the Bank will continue and expand the valuable technical and financial assistance it has rendered to the Group."

Naturally, there has been opposition and hostility to some of the Institutes' programs. Indian leaders who promoted the widespread planting of the Mexican wheats worked against strong political opposition and in fact staked their futures on their faith in CIMMYT's experts; the Green Revolution amply vindicated their steadfastness. IR8 had vociferous enemies in the Philippines in the early days of IRRI; and other examples could be cited. But the overwhelming consensus is favorable to the institutes, and nations vie with one another for the privilege of being host when the site for a new center is under discussion. Thus the institutes are seen for what they are—significant new tools to reinforce national efforts. Direct support to nations by assistance agencies would need to be expanded.

Bellagio II-V

At its April 1969 meeting, the Group agreed that their agricultural specialists should be asked to propose any specific actions which might speed the progress of nations. In February 1970 these agriculturists met (Bellagio II) and developed proposals which were considered in April by the heads of agencies (Bellagio III).

The Group agreed with their specialists that top priority should be given to adequate financing, including necessary expansion, of IRRI, CIMMYT, IITA and CIAT. They also agreed with the specialists on a number of urgent and unmet needs. First, while the nations of Asia could get help with rice, wheat and maize programs from IRRI and CIMMYT, most had no place to turn for assistance with major dryland crops such as sorghum and millets. Major food legumes still were neglected. Nations continued to build irrigation schemes without developing the cropping systems that would permit high yields to be obtained with irrigation. Intensive animal production was excluded from a major portion of sub-Saharan

Africa because of trypanosomiasis; even if trypanosomiasis were brought under control, other diseases, crude management practices, and unproductive pastures, would be among other limiting factors. There was concern for the adequacy of the knowledge base on which national agricultural planning depends.

The Bellagio Group called for intensive studies of these major problems, with reports to be in their hands by December 1970 when they would meet in New York. A sense of urgency, evident in their discussions, was reflected in the shortening of the period between meetings.

One other major development at Bellagio III was the realization that some means of marshalling substantial financial resources would be required to build the networks of research and training activities. The Group urged Mr. McNamara and the World Bank to provide leadership in the formation of an international consortium of donor agencies to consider the financial requirements which even then were estimated at \$40–50 million annually by the mid-1970s.

During the summer of 1970 the tempo picked up. Virtually all available agricultural staff members of the Ford and Rockefeller foundations and of Canada's International Development Research Centre set aside other responsibilities to undertake the several studies which had been requested by December by Bellagio III.

In December 1970, in the board rooms of the Ford and Rockefeller foundations, they received the reports which were ready. And they learned that the World Bank had approved the Bank's sponsorship—with UNDP and FAO—of an International Consultative Group for Agricultural Research.

The Consultative Group Organizes

Barely 2 months after Bellagio IV, representatives of nations, international agencies, and foundations met at the World Bank to discuss the desirability and feasibility of organization of the Consultative Group. For many of the governments the subject was new, their representatives as yet had little knowledge either of the institutes or the concept of the developing international network of scientific and technical activities. Most were at the first meeting to listen.

By May the Secretariat at the Bank had received enough indications of interest to justify the new initiative. Among those accepting membership were the World Bank, FAO and UNDP; Canada, Denmark, France, Federal Republic of Germany, Netherlands, Norway, Sweden, United Kingdom, United States; the African and Inter-American Development Banks; the IDRC and the Ford, Kellogg and Rockefeller foundations.

In many ways the Consultative Group is unique. Membership is entirely voluntary and members are under no obligation to provide funds for any activity. They do agree to listen jointly to the needs for financial support and to act in concert to the degree possible in matching funds with needs. Having heard the requirements of each organization, most donors make direct arrangements with the institutions they support. The Group Secretariat periodically makes compilations available to each donor and works with donors in fitting resources to needs.

Second, the Group bases its action exclusively on recommendations of a twelve-person Technical Advisory Committee, most of them distinguished scientists of developing countries. It informs the Group on priorities, quality of the technical programs of institutions seeking assistance, and on reasonableness of financial support sought. It is headed by Sir John Crawford of Australia with Dr. M. S. Swaminathan, Director, Indian Council of Agricultural Research, as vice chairman.

Third, the Group, comprising as it does a mixture of national, international, and private organizations, has set aside much of the protocol and red tape which characterizes so many official international organizations. It has flexibility and one senses a genuine desire on the part of most members to work together.

Fourth, the consortium of donors should be able to provide continuity to each institute and develop the network—activities that most donors could not do individually because of uncertainties of their own sources of funds.

Fifth, it involves close working arrangements between scientist and financier, donor and recipient, generalist and specialist, with action based on consensus. There are few votes taken, and few written agreements.

The Consultative Group met to consider financial needs of the four institutes for the first time in late 1971, when it successfully raised the $15 million for 1972. It also provided interim funding for a fifth institute, the new International Potato Center with headquarters in Peru.

The Launching of ICRISAT

During 1972 negotiations were completed for the establishment in India of the first international institute to be created under auspices of the Consultative Group itself. It will concentrate on improvement of sorghum and millets (cereal grains) and on chick-peas and pigeon peas (food legumes), all of which are important in the areas where rainfall is limited. Like the other institutes, it will have an international staff and an international, self-perpetuating board of trustees.

Unlike the other institutes, ICRISAT (International Crops Research Institute for the Semi-Arid Tropics) was founded by agreement of the World Bank and the United Nations Development Programme (for the Group) and the Government of India, which made available 1300 hectares of land 15 miles from Hyderabad.

Initial financial supporters of ICRISAT include Canada, the Federal Republic of Germany, Canada's International Development Research Centre, Norway, Sweden, Switzerland, the United Kingdom, UNDP, the United States, and the World Bank. The Ford Foundation served as the executing agency for the Group, its officers working closely with authorities of India. That they were able to complete the negotiations, identify and obtain a site, complete legal arrangements, and identify sources of some $12 million for capital costs and an eventual $5 million in annual operating expenses, is a tribute to all concerned. It also attests to the determination of the members of the Consultative Group to move decisively to meet the more urgent needs. For 1974 the Consultative Group appears to have raised the $30 million required, with contributions from a dozen nations, IDRC, UNDP, the World Bank and the Ford, Kellogg and Rockefeller foundations.

Recent additions to the network of cooperating centers are the International Potato Center in Peru and the Asian Vegetable Research and Development Center in Taiwan, an associate member of the group.

During 1974 it is expected that two new centers will be established. One of particular importance to the drought-afflicted sub-Saharan region is the International Livestock Center for Africa (ILCA), which is expected to conduct research through a series of regional field stations in different ecological zones. Another, the International Laboratory for Research on Animal Diseases in Kenya, will focus sharply on a search for vaccines effective against trypanosomiasis and East Coast Fever—costly, debilitating, and often fatal diseases of cattle in much of the humid tropics of Africa.

Finally, the Consultative Group and the Technical Advisory Committee are giving attention to an intensified, coordinated world program of collection, description, evaluation and preservation of crop and other germplasm, to be carried out in close cooperation with FAO, the institutes, and important national organizations on all continents.

This new world initiative has involved scores of eminent personalities of the developing countries who have served on the boards of the institutes. Among them are Virgilio Barco, Chairman of the Board of CIMMYT; Secretary of Agriculture Bernardo Aguirre of Mexico; Dr. Galo Plaza, Secretary-General of OAS; General Carlos Romulo, Foreign Minister of the Philippines; Dr. Bukar Shaib, Permanent Secretary of Agriculture of Nigeria and Chairman of the Board of IITA; Don Francisco de Sola of El Salvador and Chairman of the Board of CIAT; Prince Chakrabandhu of Thailand; Robert Gardiner of the Economic Commission for Africa; and Shri C. Subramaniam, Minister of State for Industrial Development, Science & Technology, of India. The twelve distinguished scientists on the Technical Advisory Committee, working under tremendous pressure of limited time, have nevertheless quickly placed before the Consultative Group the results of extensive studies which have been used as the basis for Group financial commitments. The World Bank and its president, Robert McNamara, moved decisively to create the Consultative Group with the strong support of Paul Hoffman of UNDP and Dr. Adekke Boerma of FAO and with the urging of members of the original Bellagio Group.

Finally, much credit for the original concept of the institutes must go to Dr. J. G. Harrar, former President of The Rockefeller Foundation, and to Dr. F. F. "Frosty" Hill of Ford who have committed much of their time over the past 10 years to assistance to nations through the institute system.

Biographical Notes

Dr. Eugene Rabinowitch

Eugene Rabinowitch was born in Russia in 1901. He studied and received his doctorate at the University of Berlin. He came to the U.S. in 1938. From 1944 to 1946 he served with the Manhattan Project, and has been the organizer of, and key participant in, the Pugwash international conferences on science and public affairs. Dr. Rabinowitch was a Professor of Biophysics at the University of Illinois from 1960 to 1968, Professor of Chemistry and Biology and director of the Center for the Study of Science and Society at the State University of New York at Albany from 1968 to 1972. He has been editor of the *Bulletin of the Atomic Scientists* since its inception in 1945 and was a fellow at the Woodrow Wilson International Center for Scholars, Smithsonian Institution, from 1972 until his death in 1973.

Dr. Victor Rabinowitch

Since receiving his Ph.D. in Zoology and International Affairs from the University of Wisconsin, Dr. Rabinowitch has spent nearly 10 years in international scientific matters at the U.S. National Academy of Sciences where he is currently Director of the Board on Science and Technology for International Development. Prior to his current assignment, he was Associate Professor of Science and Public Administration at the State University of New York at Albany, responsible for the establishment of the Center on the Problems of Science and Society. Dr. Rabinowitch is currently a member of the U.S. Committee for Pugwash, the Convenor of the Pugwash Study Group on Science and Technology in Development and Vice President of the Institute on Man and Science in Rensselaerville, New York. He is a member of the American Academy of Arts and Sciences' Committee on Developing Nations, the Committee on Science and Public Policy of AAAS, a Director and Editorial Consultant of the *Bulletin of the Atomic Scientists* and a member of the Editorial Boards of *Public Science*, *World Development* and *Science Studies*.

Dr. Hyung Sup-Choi

Before Dr. Choi received his M.S. in physical metallurgy from the University of Notre Dame and his Ph.D. in metallurgical engineering from the University of Minnesota in 1958, he was a metallurgical engineer in industry in Korea and also taught at Seoul National University and the Korean Naval Academy. After his return from the United States he became a senior research scientist at the Atomic Energy Research Institute of Korea of which he became Director in 1964. This was followed by another important position in scientific research administration in 1966 when he became the first president of the Korea Institute of Science and Technology, considered a unique example of an industrial research institute in a developing country. He was responsible for its original organization during the 5 years he was with the Institute and was concurrently a member of the President's Council of Economic and Scientific Advisors. In June of 1971 Dr. Choi was appointed Minister of Science and Technology of the Republic of Korea, a post which he now holds. Dr. Choi has contributed over forty papers to professional publications such as the *Transactions of the American Institute of Mining, Metallurgical, and Petroleum Engineers*, the *Journal of Applied Physics*, the *Journal of the Korean Chemical Society*, the *Canadian Metallurgical Quarterly*, and the *Journal of the Mining and Metallurgical Institute of Japan*, and he holds a dozen patents.

Dr. Jose Barzelatto

Jose Barzelatto, M.D. is presently special advisor to the Chairman of the Permanent Executive Committee of the Interamerican Council for Education, Science and Culture (OAS) and member of the Continuing Committee of the Pugwash Conference on Science and World Affairs. A Chilean citizen, he is former Associate Professor of Medicine at the University of Chile and Executive Secretary of the Chilean National Council for Science and Technology and has served as a consultant to IAEA, WHO/PAHO and ECOSOC.

Professor L. K. H. Goma

Professor Lameck Kazembe Haza Goma, who received his B.A. and M.A. from Queen's College, University of Cambridge, and his Ph.D. from the University of London, has been actively involved in scientific education and research in Africa for the past 20 years. Currently he is Vice-Chancellor of the University of Zambia, Chairman of the National Food and Nutrition Commission and Vice-Chairman of the National Council for Scientific Research in Zambia. In addition he is Chairman of the Scientific Council of Africa, Organization of African Unity; a member of the Executive Board of the Association of African Universities and a member of the Council of the Association of Commonwealth Universities. He has twenty-seven scientific publications to his credit including papers presented at international conferences, and two books.

Dr. A. Parthasarathi

Ashok Parthasarathi is currently Special Assistant for Science and Technology in the Prime Minister's Secretariat in India, and is a physicist and electronics engineer by training. He has worked at the Mullard Radio Astronomy Observatory of Cambridge University, U.K., has taught physics at the Birla Institute of Technology and Science, Pilani, and researched at the Massachusetts Institute of Technology, U.S.A. on science and public policy. Joining the Department of Atomic Energy in 1967, he was instrumental in setting up a Programme Analysis Group as a policy planning and management services unit for the Atomic Energy Commission. Concurrently he worked on the staff of the former Electronics Committee of the Government of India. He has published a number of papers on various aspects of science and technology policy, planning and management, and has directed a multi-institutional study of the Scientific Instrument Industry, which is currently in press.

Mr. Parthasarathi was a member of the Indian Delegation to the U.N. Conference on the Peaceful Uses of Outer Space in 1968 and the U.N. Conference on the Human Environment in 1972. He has represented India on the inter-governmental Committee on the Transfer of Technology of the UNCTAD, and has also been a member of a number of U.N. Expert Panels on various Science and Technology policy issues. He is Executive Secretary of the Indian Pugwash Society and has attended a number of Pugwash General Conferences and Symposia.

Dr. Francisco Sagasti

Dr. Francisco Sagasti was born in Peru and received his professional engineer's degree from the National Engineering University in Lima. He received his M.Sc. in Industrial Engineering from Pennsylvania State University and Ph.D. in the Philosophy of Operations Research from the University of Pennsylvania. Following his graduation he was Senior Specialist in the Department of Scientific Affairs of the Organization of American States, working with the Secretariat of the Andean Common Market in Lima. Currently he is associated with the Science and Technology Instrument Project in Lima. Dr. Sagasti has served as a consultant to international institutions including the Interamerican Development Bank, The Department of Scientific Affairs of the Organization of American States, as well as to organizations within Peru. He is also the author of a number of publications.

Dr. Carlos M. Varsavsky

Carlos Varsavsky was born in Buenos Aires, Argentina, in 1933 and studied engineering physics at the University of Colorado and astronomy at Harvard, taking his Ph.D. there in 1958. He has had considerable teaching experience in North and South America and is currently Professor of Astronomy at the University of La Plata, Argentina. He has held many research positions and appointments in industry, including Technical Directorship of the ALUAR Aluminio Argentino SAIC, currently he is Manager for Research and Development of FATE in Buenos Aires. Dr. Varsavsky has thirty-four scientific contributions to international journals and encyclopaedias to his credit and four books.

Professor Yash Pal

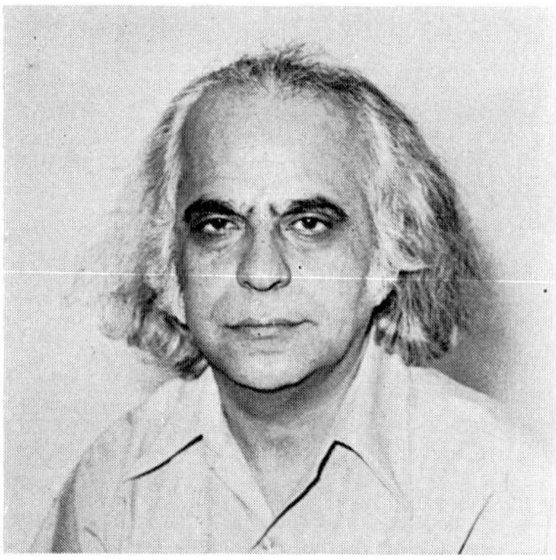

Professor Yash Pal is the Director of the Space Applications Centre at Ahmedabad, and Professor of Physics at the Tata Institute of Fundamental Research, Bombay. Professor Pal was born at Jhang, Pakistan, in 1926, obtained his Bachelor's and Master's degrees in India at the Punjab University, Lahore and Delhi, and did his doctoral work at the Massachusetts Institute of Technology. He has worked extensively in areas of Cosmic Ray Physics, High Energy Physics and Theoretical Astrophysics.

Professor Pal has been active in India in problems relating to science and society, has helped initiate a programme of curriculum development in School Science, both for urban and rural environment of India. He is currently directing an institute (Space Applications Centre) which is engaged in a major techno-social experiment to bring satellite television, with developmental messages, to several thousand remote Indian villages.

Dr. Thomas R. Odhiambo

Thomas R. Odhiambo received his M.A. and Ph.D. from the University of Cambridge, U.K., and is currently Director of the International Centre of Insect Physiology and Ecology in Nairobi, Kenya. Dr. Odhiambo has been very active in scientific research activities in East Africa. He is the founder of the East Africa Academy and has attended a number of international conferences. He has over eighty-five publications to his credit including papers presented at international conferences.

Dr. A. R. Abdel Meguid

Dr. A. R. Abdel Meguid, a Pugwash member, is currently Secretary of State in the Prime Minister's office in Egypt. He was senior economist with the World Bank (1970–4) and U.N. Chief Advisor to the Government of Ceylon. Previously he was Director General of the Asswan Regional Planning Authority.

Dr. Frank G. Torto

Frank G. Torto is Professor and Head of the Department of Chemistry, University of Ghana, Legon. He is a past Fellow of Churchill College, Cambridge, and a former consultant on Science and Technology at the United Nations in Geneva. He has served on many national and international bodies concerned with education and with science policy and development, and has written and lectured on these themes. He is currently a member of the Continuing Committee of the Pugwash Conferences on Science and World Affairs.

Dr. W. K. Chagula

The Hon. Dr. W. K. Chagula, M.B., Ch.B. (E.A.), M.A. (Camb.), Fellow of E. A. Academy, is President of The East African Academy and Vice-Chairman of the International Centre of Insect Physiology and Ecology. He is also a member of the United Nations Advisory Committee on the Application of Science and Technology to Development, and a former member of the Pugwash Continuing Committee.

Dr. Chagula has worked as Assistant Medical Officer in Tanganyika (1952), Lecturer in Anatomy, Makerere University College, Uganda (1955–62), Registrar and Vice-Principal of The University College, Dar es Salaam, Tanzania (1963–5), Principal of The University College, Dar es Salaam (1965–70), Tanzania Minister for Water Development and Power (1970–2), and is currently Tanzania Minister for Economic Affairs and Development Planning.

Dr. Harrison Brown

Harrison Brown was born in Wyoming in 1917 and attended the University of California at Berkeley. He took his Ph.D. in chemistry at the Johns Hopkins University in 1941 and went from there to the Manhattan Project. He was Assistant Director of Chemistry at the Clinton Laboratories in Oak Ridge from 1943 to 1946. After World War II he joined the faculty of the Institute for Nuclear Studies at the University of Chicago. In 1951 he became Professor of Geochemistry at the California Institute of Technology. In 1968 he was also appointed Professor of Science and Government. He served as Foreign Secretary of the U.S. National Academy of Sciences from 1962–1974.

Dr. Carl Djerassi

Carl Djerassi is Professor of Chemistry at Stanford University as well as President of Zoecon Corporation —a developer of new methods of insect control. He is a member of the U.S. National Academy of Sciences, the American Academy of Arts and Sciences and a foreign member of the Royal Swedish Academy of Sciences and the German Academy, Leopoldina. He is Chairman of the National Academy's Board on Science and Technology for International Development and he has been active for many years in development problems in the Pugwash Movement where he is currently serving as a member of the Continuing Committee. In 1973 he was awarded the National Medal of Science for his work on oral contraceptives.

Dr. Herbert Marcovich

Dr. Herbert Marcovich is currently Director of Research at the National Center for Scientific Research, University of Paris. He was born in Cairo and is a doctor of medicine as well as a doctor of sciences. His main fields of activity have included bacterial genetics and radiobiology. Dr. Marcovich has been active in the Pugwash movement since 1958 and served on the Continuing Committee from 1962 to 1972.

Dr. Stevan Dedijer

Dr. Stevan Dedijer is currently Professor of Sociology of Science and founder and present director of the Research Policy Program at Lund University in Sweden. Born in Yugoslavia, Dr. Dedijer received his degree in Mathematical Sciences from Princeton University. Prior to his current position, he was a visiting fellow at the Center for Advanced Study in the Behavioral Sciences at Stanford University and a visiting professor, Department of the History on Sciences at Yale University and Department of Government, Harvard University. He has been a consultant and/or advisor on research and development policy to various scientific organizations such as Yugoslav Atomic Energy Commission, Yugoslav National Research Council, Venezuelan National Research Council, OECD, Indian Planning and Education Commissions, and the Council of Scientific and Industrial Research.

Dr. Charles Weiss

Charles Weiss, a United States citizen, was born in San Francisco and grew up in Philadelphia. He graduated, *summa cum laude* in chemistry and physics, from Harvard College in 1959, and earned his Ph.D. in chemical physics and biochemistry from Harvard University in 1964. He then did research in biophysics and theoretical chemistry at the Night Vision Laboratory, the Lawrence Berkeley Laboratory at the University of California, and the IBM Watson Laboratory at Columbia University. He is now Science and Technology Adviser to the International Bank for Reconstruction and Development (the World Bank). The present article was written while Dr. Weiss was at IBM.

Dr. Lawrence W. Bass

Lawrence W. Bass, a graduate from Yale University, where he received a Ph.D. in 1922, did post-doctoral research at the Sorbonne, Pasteur Institute, and Rockefeller Institute. He has had a 50-year professional career of fundamental research on nucleic acids and proteins, 15 years as a technical director in industry, and more than 25 years as a contract research executive at the Mellon Institute and Arthur D. Little Inc. He officially retired in 1964 and is now the author of over 150 publications, including four books and the basic manuscripts for four UNIDO monographs on technical management in developing countries.

Mr. W. Murray Todd

W. Murray Todd received his B.A. in political science from Stanford University and M.B.A. from the Stanford Graduate School of Business. After spending several years in private industry with the Levi Strauss Company in San Francisco he was involved in international affairs with the U.S. Government. Since 1962 he has been Executive Secretary of the Office of the Foreign Secretary in the U.S. National Academy of Sciences. In this capacity he has been involved with the coordination of the Academy's cooperation with sister scientific institutions abroad.

Dr. Roger Revelle

In three decades, the scientific explorations of Roger Revelle have covered a remarkable range of activities and have opened up large tracts of scientific terrain for experimenters and theoreticians. Dr. Revelle, who received his Ph.D. from the University of California in oceanography, is a former Director of the Scripps Institution. During the time he held that position, he was instrumental in the creation of the San Diego campus of the University of California. He also led a number of expeditions into the deep Pacific that contributed to knowledge of oceanic processes and the geology of the sea floor. Dr. Revelle promoted international co-operation in oceanography and participated in a number of intergovernmental delegations. From 1964 to 1967 he was President of the International Association of Physical Sciences of the Ocean. In addition he was appointed the first chairman of the U.S. National Committee for the International Biological Program. Revelle became science adviser to the Secretary of the Interior in 1961 and turned to an entirely new set of problems relating to world population growth and economic transformation. His concern with the population problem led him to accept the Richard Saltonstall Chair of Population Policy and the directorship of the Harvard Center for Population Studies. For the last 10 years his primary interest in international scientific co-operation has been in the use of science and technology to assist the less-developed countries. For several years he has been deputy foreign secretary of the National Academy of Sciences with primary responsibilities in problems of science and technology for the developing nations.

Dr. Vikram A. Sarabhai

Dr. Vikram Ambalah Sarabhai, the father of India's space programme, has left his impress on varied fields of human endeavour. Born in Ahmedabad, India, in 1919, he received his doctorate from Cambridge University. In 1962 Dr. Sarabhai took over the responsibility of organizing space research in India and became the Chairman of the Indian National Committee for Space Research for setting up the Thumba Equatorial Rocket-launching Station and initiated a programme for the manufacture of French Centaure sounding rockets in India. He saw the tremendous potential of using modern science to help in the process of development. He drew up plans to take modern education to the remotest villages by using satellite television. His interest in education resulted in his establishing the Community Science Centre where new ideas in science education could be tried. Dr. Sarabhai died in 1971.

Dr. Theodore W. Schultz

Theodore W. Schultz received his Ph.D. from the University of Wisconsin in 1930. He was a member of the Department of Economics faculty at Iowa State during 1930–43. Since then he has been at the University of Chicago, serving as Chairman of the Department of Economics at that University from 1946 to 1961. He was President of the American Economic Association in 1960 and is a founding member of the National Academy of Education. He received the Francis A. Walker award for original contributions to economics, 1972. He is a fellow of the American Academy of Arts and Sciences, of the American Philosophical Society, of the American Agricultural Economics Association, and of the Center for Advanced Study in Behavioral Sciences. He is a long-time Trustee for the Population Council, and presently, a governor of the Board of the International Economic Development Center of Canada. His major professional studies are in Agriculture, Investment in Research, Human Capital and Investment in Education.

Dr. Sterling Wortman

Dr. Wortman, now Vice-President, The Rockefeller Foundation, New York, served for 5 years as director for Agricultural Sciences of the Foundation with responsibilities for agricultural activities worldwide. Following receipt of the Ph.D. degree in plant genetics at the University of Minnesota in 1950, he headed corn breeding activities in Mexico and Central America for the Foundation until 1955 when he became head of the plant breeding department at the Pineapple Research Institute of Hawaii. He then served as associate director of the International Rice Research Institute in the Philippines during its formative years, after which he served briefly as director of the Pineapple Research Institute before moving to the Foundation's New York office. He has been active in the establishment of the International Maize and Wheat Improvement Center (CIMMYT, from the Spanish) and serves as vice chairman of the board. He negotiated with the government of Colombia for the establishment of the International Center of Tropical Agriculture near Cali. He has been directly involved in many of the activities which have led to the formation of the Consultative Group for International Agricultural Research.

Dr. Wortman is a member of the Board on Science and Technology for International Development of the U.S. National Academy of Sciences; a member of the Research Advisory Committee, National Science Foundation; and a member of the Advisory Panel on Agricultural and Rural Development, World Bank.

Dr. Alvin M. Weinberg

Alvin M. Weinberg is Director of the Office of Energy Research and Development (OERD), which is part of the Federal Energy Office and of the Executive Office of the President. He is also Chairman of the newly established governmental Inter-agency Energy Research and Development Committee. Dr. Weinberg began his research career as a mathematical biophysicist. In early 1945 he became one of the first members of the University of Chicago's war-time Metallurgical Laboratory, where he helped design the first large nuclear power reactors. For his role in the development of nuclear reactors, Weinberg in 1960 shared the Atoms for Peace Award and was one of the first recipients of the E. O. Lawrence Memorial Award for his contributions to nuclear reactor theory and for work in the design of production, research, and power reactors. He has also been awarded The University of Chicago Alumni Medal, The University of Tennessee N.W. Dougherty Award in Engineering, and the American Nuclear Society's Jack Chernick Memorial Award.

Dr. Weinberg has a number of academic connections and honorary degrees and serves on the Advisory Boards of several learned journals. In 1969–70 he served on President Nixon's Task Force on Science Policy, and in 1972 was a member of the National Cancer Plan Evaluation Committee. He is a member of the American Academy of Arts and Sciences, a Fellow of the American Nuclear Society, Fellow of the American Physical Society where he served on the Council from 1955 to 1959, a member of the American Association for the Advancement of Science and a member of the National Academy of Sciences (NAS). He was Vice-Chairman of the NAS Committee which prepared the report *Applied Science and Technological Progress* and served as a member of the Council of the National Academy from 1967 to 1970.

Dr. Ignacy Sachs

Ignacy Sachs was born in 1927 in Warsaw and is French by naturalization. He studied economics in Rio de Janeiro, Delhi (Ph.D., 1961), and Warsaw (Doc.hab., 1962).

From 1961 to 1968 he was assistant and then associate professor at the Main School of Planning and Statistics, and Director of the Center of Research on Underdeveloped Economies, sponsored by the Warsaw University and the School of Planning and Statistics.

Since November 1968 he has been Director of Studies at the École Pratique des Hautes Études, VI° Section, Paris, and Director of its International Research Center on Environment and Development.

Dr. Norman E. Borlaug

Norman E. Borlaug is Director of the Wheat, Barley and Triticale Research and Production Programs at the International Center for Maize and Wheat Improvement (CIMMYT). Born in Cresco, Iowa, in 1914, he first entered the U.S. Forest Service in 1935. From 1942 to 1944 he was a microbiologist with E. I. Dupont de Nemours. After 16 years as a research scientist in charge of Wheat Improvement in Co-operative Mexican Agricultural Program he became Associate Director, The Rockefeller Foundation, assigned to Inter-American Food Crop Program from 1960 to 1963, and joined CIMMYT in 1964. Dr. Borlaug has held numerous consultancies on wheat research and production problems and in this capacity has served many governments in developing countries. He has been a member of the Citizens' Commission on Science, Law and Food Supply and of the Commission on Critical Choices for America since 1973.

Dr. Borlaug has received more than sixty academic, scientific, government, civic and achievement awards which include the Nobel Prize for Peace in 1970. As well as being a member of the U.S. National Academy of Sciences since 1968, he was made an Honorary Foreign Academician of N.I. Vavilovi Academy of Agricultural Sciences of The Lenin Order (U.S.S.R.) in 1972, and a Foreign Member Indian National Science Academy (India) in 1973.

The Sussex Group

Early in 1968 the secretariat of the United Nations Advisory Committee on the Application of Science and Technology for Development commissioned a group of natural and social scientists at the University to prepare a draft introductory statement to the U.N. Plan of Action on the Application of Science and Technology to Development. The Sussex Group consisted of Mr. Charles Cooper, Dr. Oscar Gish, Professor Christopher Freeman, Dr. Stephen Hill, Dr. Geoffrey Oldham and Professor Hans Singer.

All of the group were associated with either the Science Policy Research Unit or the Institute of Development at the University of Sussex. They met for one week at the University during the month of July 1968 and were joined by Mr. Raj Desai from the ACAST Secretariat.

Two members of the group (C. Freeman and C. H. G. Oldham) attended the meeting of the Advisory Committee which was held in Addis in November 1968, and participated in a lively debate on the draft chapter. After intensive discussion it was decided that the Sussex Manifesto would not be used as the introductory chapter but rather should be published as a part of a more general statement of the United Nations Advisory Committee's views on the role of Science and Development. This led to the publication, in 1970, of the United Nations report entitled *Science and Technology for Development: Proposals for the Second United Nations Development Decade*, Report of the Advisory Committee on the Application of Science and Technology to Development (ST/ECA/133).